DECONSTRUCTING COSMOLOGY

The advent of sensitive high-resolution observations of the cosmic microwave background radiation and their successful interpretation in terms of the standard cosmological model has led to great confidence in this model's reality. The prevailing attitude is that we now understand the Universe and need only work out the details. In this book, Sanders traces the development and successes of Lambda-CDM, and argues that this triumphalism may be premature. The model's two major components, dark energy and dark matter, have the character of the pre-twentieth-century luminiferous aether. While there is astronomical evidence for these hypothetical fluids, their enigmatic properties call into question our assumptions of the universality of locally determined physical law. Sanders explains how modified Newtonian dynamics (MOND) is a significant challenge for cold dark matter. Overall, the message is hopeful: the field of cosmology has not become frozen, and there is much fundamental work ahead for tomorrow's cosmologists.

ROBERT H. SANDERS is Professor Emeritus at the Kapteyn Astronomical Institute of the University of Groningen, the Netherlands. He received his Ph.D. in astrophysics from Princeton University under the supervision of Lyman Spitzer. After working at Columbia University and the National Radio Astronomy Observatory, he moved to Europe. He spent his career studying active galactic nuclei (in particular, the Galactic Center), on the hydrodynamics of gas in galaxies, and, for several decades, on the problem of the "missing mass" in astronomical systems. His previous books are *The Dark Matter Problem: A Historical Perspective* (2010) and *Revealing the Heart of the Galaxy: The Milky Way and Its Black Hole* (2013).

DECONSTRUCTING COSMOLOGY

ROBERT H. SANDERS

Kapteyn Astronomical Institute, The Netherlands

CAMBRIDGE
UNIVERSITY PRESS

University Printing House, Cambridge CB2 8BS, United Kingdom

Cambridge University Press is part of the University of Cambridge.

It furthers the University's mission by disseminating knowledge in the pursuit of education, learning and research at the highest international levels of excellence.

www.cambridge.org
Information on this title: www.cambridge.org/9781107155268

© Robert H. Sanders 2016

This publication is in copyright. Subject to statutory exception and to the provisions of relevant collective licensing agreements, no reproduction of any part may take place without the written permission of Cambridge University Press.

First published 2016

Printed in the United Kingdom by TJ International Ltd. Padstow Cornwall

A catalogue record for this publication is available from the British Library

Library of Congress Cataloguing in Publication data
Names: Sanders, Robert H.
Title: Deconstructing cosmology / Robert H. Sanders.
Description: Cambridge : Cambridge University Press, 2016. |
Includes bibliographical references and index.
Identifiers: LCCN 2016026392 | ISBN 9781107155268 (hardback : alk. paper)
Subjects: LCSH: Cosmology. | Dark matter (Astronomy) |
Cosmic background radiation. | Newtonian cosmology.
Classification: LCC QB981.S318 2016 | DDC 523.1–dc23
LC record available at https://lccn.loc.gov/2016026392

ISBN 978-1-107-15526-8 Hardback

Cambridge University Press has no responsibility for the persistence or accuracy of URLs for external or third-party internet websites referred to in this publication, and does not guarantee that any content on such websites is, or will remain, accurate or appropriate.

Contents

Acknowledgments		*page* vii
Introduction		1
1	Creation Mythology	6
2	Three Predictions of Physical Cosmology	12
	2.1 The Basis of Physical Cosmology	12
	2.2 The Expansion of the Universe	14
	2.3 The Background Radiation	18
	2.4 Anisotropies in the Background Radiation	23
3	The Very Early Universe: Inflation	28
	3.1 Fine-Tuning Dilemmas and the Initial Singularity	28
	3.2 An Early De Sitter Phase	30
	3.3 The Physical Basis of Inflation	32
4	Precision Cosmology	37
	4.1 Standard CDM Cosmology	37
	4.2 Primordial Sound Waves	41
	4.3 The ΛCDM Paradigm	47
5	The Concordance Model	49
	5.1 Consistency	49
	5.2 *WMAP* and *Planck*	51
	5.3 The Density of Baryons	53
	5.4 Supernova Cosmology	55
	5.5 Hubble Trouble?	57
	5.6 Baryon Acoustic Oscillations	58

	5.7 The "Axis of Evil"	61
	5.8 Summing Up: Absence of Discord	63
6	Dark Energy	66
	6.1 The Evidence for Dark Energy	66
	6.2 The Nature of Dark Energy	70
	6.2.1 Zero-Point Energy	70
	6.2.2 Dynamic Dark Energy	71
	6.3 Dark Energy and Fundamental Physics	75
7	Dark Matter	77
	7.1 Evidence for Dark Matter in Galaxies and Galaxy Systems	77
	7.2 Cosmological Evidence for Dark Matter	81
	7.3 The Nature of Dark Matter	82
	7.4 The Science of Dark Matter Detection	85
	7.4.1 Indirect Detection of Dark Matter	85
	7.4.2 Direct Detection of Dark Matter	89
	7.4.3 The LHC and Dark Matter	91
	7.5 The Sociology of Dark Matter Detection	93
8	MOND	96
	8.1 Galaxy Phenomenology Reveals a Symmetry Principle	96
	8.2 An Empirically Based Algorithm	101
	8.2.1 Galaxy Rotation Curves	101
	8.2.2 The Baryonic Tully–Fisher Relationship	104
	8.2.3 A Critical Surface Density	107
	8.3 Cosmology and the Critical Acceleration	110
	8.4 Problems with MOND	111
9	Dark Matter, MOND and Cosmology	115
	9.1 The Puzzle	115
	9.2 Particle Cosmic Dark Matter	117
	9.2.1 Neutrinos	117
	9.2.2 Soft Bosons	119
	9.3 New Physics	120
	9.4 Reflections	122
10	Plato's Cave Revisited	126
Notes		130
Index		141

Acknowledgments

I express my gratitude first of all to my old friend Frank Heynick of Brooklyn, NY. Frank spent some time and effort in reading an initial draft of this book and giving me his very useful advice as an intelligent, interested non-expert. This led to considerable improvements to the content and readability. I also thank Phillip Helbig for a later reading of the complete manuscript. Not only did Phillip (with his eagle eyes) spot my many typographical and grammatical errors, but he also made substantial comments and criticisms on scientific and philosophical issues.

I have had numerous useful conversations with my colleague Saleem Zaroubi on a number of aspects of the cosmological paradigm. It was very beneficial to have the highly informed opinions of a thoughtful supporter of the standard model. Thanks to Saleem, I also benefitted from a short but useful conversation with Naoshi Sugiyama on the phase focussing of primordial sound waves (the reason we see distinct acoustic peaks in the CMB angular power spectrum) and its consistency with the inflationary paradigm. I also gratefully acknowledge a helpful conversation with Alexi Starobinsky over the early history of the idea of cosmic inflation as it developed in the Soviet Union.

I thank Moti Milgrom for his extensive and very helpful comments on the MOND chapters. Moti's characteristically deep and insightful criticisms greatly improved the presentation and gave me a new appreciation and understanding of the significant space–time scale invariance of the deep MOND limit.

I am very grateful to Stacy McGaugh for making figures available in legible form. Stacy deserves much credit for his persistent emphasis over the years on the beautiful simplicity and importance of the baryonic Tully–Fisher relation as a test of CDM and of MOND.

I wish to acknowledge the many conversations, letters and emails over 30 years with Jacob Bekenstein. Jacob did not have the chance to read over this manuscript but he certainly influenced the content and range of this book. His profound understanding of physics in general and relativity in particular benefitted all who had the privilege of knowing and interacting with him. His explanations were clear and intuitive (in a subject that is not so intuitive), and with me he was always patient and considerate. He will be greatly missed.

And finally I thank Vince Higgs of Cambridge University Press for his help, guidance and advice throughout this project.

Introduction

Under the sub-subject of cosmology, Amazon.com currently lists 5765 items. Among them there are textbooks, serious scientific discussions, popular books, books on history, philosophy, metaphysics and pseudoscience, mega-bestsellers like those by Stephen Hawking, Brian Greene and Lisa Randall, and works no one has heard of by authors as obscure as their books. It would almost seem as though the number of books on the subject is expanding faster than the Universe; that soon the nature of the missing mass will be no mystery – the dark matter is in the form of published but largely unread cosmology books. Does the world need yet another book about this subject? Why have I decided to contribute to this obvious glut on the book market? Why do I feel that I have something to add of unique value?

The idea for the current project had its dim origins in the year 2003 when I was invited to lecture on observational cosmology at a summer school on the Aegean island of Syros. I was surprised at this invitation because I am neither an observer nor a cosmologist; I have always worked on smaller-scale astrophysical problems that I considered soluble. In this career choice I was no doubt influenced by my first teachers in astrophysics, who were excellent but traditional and, to my perception at least, found cosmology to be rather fanciful and speculative (although I never heard them explicitly say so and almost certainly they would not say so now).

But I decided that this invitation was an opportunity to learn something new, so I prepared a talk on the standard cosmological tests (e.g., the Hubble diagram, the angular size–redshift relation and the number counts of faint galaxies) in the context of the current cosmological paradigm that is supported by modern observations, such as the very detailed views of tiny anisotropies in the cosmic microwave background radiation (CMB). The issue I considered was the overall consistency of these classical tests with the standard model – Lambda-CDM (ΛCDM).

I was actually more interested in finding inconsistency rather than consistency. This is because of my somewhat rebellious nature, as well as my conviction that science primarily proceeds through contradiction and conflict rather than through agreement and "concordance." However, somewhat to my chagrin, I found that the classical tests were entirely consistent with the standard paradigm, although with much lower precision than that of the modern CMB observations. There is no conspicuous inconsistency between the model and those observations considered to be cosmological.

And yet worries persist (not only by me) about Lambda-CDM, not because of any direct observational contradiction on a cosmological scale but because of the unknown nature of the two dominant constituents of the world – dark energy and dark matter. These two media are deemed necessary because the Universe, and more significantly, objects within it, do not behave as expected if the laws of physics on the grand scale take on the same form as that established locally. The universality of terrestrial and Solar System physical law deduced from local phenomena is an assumption, and one that is difficult to test. The required existence of these two invisible media – aethers, undetectable apart from their dynamical or gravitational effects – is perhaps a hint that the assumption of the universal validity of local physics may not be valid.

These two components – dark energy, designated by Λ, and dark matter, abbreviated as "CDM" (for cold dark matter) – comprise 95% of the energy density of the Universe, and yet we have no clear idea of what they are. Dark energy is elusive; the only evidence is, and very possibly can only be, astronomical. Is this simply a cosmological constant in Einstein's equations, or is it a vacuum energy density – the zero-point energy of a quantum field? Is it the evolving energy density of a background field – a field that is not usually included in the formulation of general relativity? Or does apparent requirement of this medium signal the breakdown of general relativity itself on a cosmological scale?

Dark matter appears to be required to explain the details of the observed temperature fluctuations in the cosmic microwave background radiation and, more fundamentally, the formation of structure in an expanding universe of finite lifetime. This same dark matter presumably clusters on the scale of galaxies and accounts for the total dynamical mass of these systems – a mass that often exceeds the detectable baryonic mass by a large factor. There is the additional problem that the dark matter particles, which should be detectable locally, have so far not made an appearance in various increasingly sensitive experiments designed to catch them, but this, I have argued, is not (and cannot be) a falsification.[1]

For the two components taken together, there is a further complication of naturalness: the density of these two "fluids" decreases differently as the Universe expands (they have different equations of state). The dark matter dilutes with the expanding volume element; the dark energy does not dilute at all (or at least very differently) with expansion. Why should these two components with different equations of state have, at the present epoch, after expansion by many factors of ten, energy densities that are comparable? This may well indicate that the models forced upon the observations are inappropriate.

There are reasonable arguments that the "mysteries" surrounding these two unknown constituents of the Universe are not really so profound and do not indicate a failure of the cosmological paradigm or of general relativity. The postulates of Einstein's general theory do permit a cosmological constant, and its phenomenological impact is the observed accelerated expansion of the Universe. The small magnitude of the constant with respect to the expectations of quantum physics does not reflect an incompleteness of general relativity but rather a lack of understanding of the nature of quantum fields in curved space. The near coincidence of the densities of matter and dark energy is not so surprising and will occur over a large range of cosmological time.[2] The dark matter particles will eventually be discovered, revealing their identity and nature.

In my opinion, the essential problem with the paradigm is that cosmology, via dark matter, impinges directly upon the dynamics of well-studied local systems – galaxies – and here, I will argue, the cosmological paradigm fails. The most profound problem, specifically with CDM, is the existence of a simple algorithm, with one new physical constant having units of acceleration, that allows the distribution of force in an astronomical object to be predicted from the observed distribution of observable matter – of baryons. These predictions are surprisingly accurate for spiral galaxies, as evidenced by the matching of observed rotation curves where even details are reproduced. This demonstrates that the total force precisely reflects the distribution of observable matter even in the presence of a large discrepancy between the traditional (Newtonian) dynamical mass and the observable mass. In the context of dark matter this fact would seem to imply a very close interaction between the dark and detectable matter, an interaction totally at odds with the nature of dark matter as it is perceived to be – a dissipationless fluid that interacts with normal matter primarily by gravity.

That algorithm, of course, is modified Newtonian dynamics (MOND), proposed more than thirty years ago by Mordehai Milgrom, and its success constitutes the greatest challenge for dark matter that clusters on the scale of galaxies.[3] In so far as such dark matter is an essential aspect of ΛCDM, MOND directly challenges the prevailing paradigm, and that becomes more than a pure scientific problem; it is a significant sociological issue. This is because ΛCDM has become something of an official religion, outside of which there is no salvation and beyond which there is only damnation. The entire phenomenology of galaxies that is described so well by MOND is systematically dismissed or relegated to the category of problems for the future. This reveals a danger to the creative process presented by a dogma that is too widely and too deeply accepted. However, here I wish to discuss not only this danger, but more positive possibilities for physics and cosmology, revealed by the perceived successes of both the ΛCDM and MOND paradigms. This, to answer the question that I posed above, is the reason that I began this project.

First of all, I would like to state clearly what this book is not. It is not a critique of modern science or the application of the scientific method to the study of cosmology. At present, there are many such absurd criticisms of the methods and results of science – so much so that one can identify an anti-rationalist or even anti-realist sentiment that is dangerous in modern society. I do not wish to be identified with those who deny many of the important results of modern physics as these relate to cosmology – from relativity to particle physics to the overall coherence of the current cosmological scenario.

The title of the book might suggest to some that from the onset I have a negative attitude toward the study of cosmology. This is not what I wish to imply by use of the term "deconstructing." Following the second definition given by the Merriam-Webster dictionary, to deconstruct is to take something apart, in this case the standard paradigm, in order to reveal possible flaws, biases or inconsistencies, and that is what I wish to achieve. Overall, I believe the reader will find that I am positive about the application of the scientific method to the study of the entire Universe; the achievements of humans in the past century in understanding the structure of the world on the grand scale are truly remarkable. But there remain, at the very least, empirical problems for the observed structure and kinematics of bound astronomical objects, primarily galaxies. I will criticize these aspects of the standard cosmological paradigm as well as the general sentiment that there is no room left for doubt.

I begin with a comparison of the standard Big Bang model to early human creation mythologies, emphasizing those aspects that are similar and those that are different. This leads inevitably to a general discussion of the philosophical issues raised by cosmology as a science and the ways in which the subject differs from the other physical sciences. After this introduction, there follows a historical interlude on the origins of Big Bang cosmology, its predictive basis and its remarkable success on an empirical level. Included here is a chapter on the proposed phenomenon of "inflation" in the very early Universe – the motivation of the idea as well as its elevation to an essential aspect of the cosmological paradigm in spite of the absence of a clear microphysical basis. Going beyond the essential Big Bang, I discuss "precision" cosmology and the "concordance" model – ΛCDM – emphasizing its consistency in explaining the observed anisotropies in the cosmic microwave background, the large-scale distribution of galaxies and the accelerated expansion of the Universe.

Finally I deal with what I see as the principal problem of the paradigm: the mysterious nature of the two dominant components – dark energy and dark matter.

In particular I dwell upon the failures of cold dark matter in explaining the basic photometric and kinematic observations of galaxies. I discuss at length the leading alternative to dark matter – modified Newtonian dynamics (MOND), its observational success and the challenge to the cosmological paradigm presented by this simple algorithm. I consider the possible modifications of physics required by MOND as well as the perceived problems of MOND with respect to cosmology. I discuss possible reconciliation of MOND on galaxy scales with the evidence for dark matter on a cosmological scale.

In conclusion (and here and there throughout) I delve into philosophical issues. But I am not a philosopher and have never studied philosophy (apart from a survey course in my second year of college – long ago). So for me, any discussion of philosophy is to enter the realm of dangerous speculation. In general this is a risky business for a scientist. On the other hand, it is difficult to avoid philosophy in any discussion of cosmology.

I have tried to omit serious mathematics throughout – essentially no formulae – but I do use standard scientific notation for the very large and small numbers encountered in astronomy and cosmology: 1.38×10^{10} years for the age of the Universe (instead of 13 800 000 000) or 10^{-32} seconds for the duration of the early inflationary period (instead of 0. [followed by 31 zeros] 1).

I also describe the mass of sub-atomic particles in energy units (via Einstein's famous formula $E = Mc^2$). As is usual, the preferred unit is the electron volt or eV ($= 1.6 \times 10^{-12}$ ergs). And of course there are keV (kilo-electron volts or one thousand eV), MeV (mega-electron volts or one million eV) and GeV (giga-electron volts or one billion eV).

For units of distance I typically use light years – the distance light travels in one year, or about 10^{18} centimeters. And of course there are kilo-light years for galactic scale distances and mega-light years for cosmic scales. In places – typically in published figures – I revert to the more astronomical distance units of parsecs (about three light years). And then there are kiloparsecs (kpc) and megaparsecs (Mpc).

Finally, a word on terminology. I use the word "world" generally to be synonymous with Universe or Cosmos. But when discussing the "multiverse," world will refer to not just our piece of it, but to the whole "shebang."

1
Creation Mythology

> When the sky above was not named,
> And the earth beneath did not yet bear a name,
> And the primeval Apsû, who begat them,
> And chaos, Tiamat, the mother of them both,
> Their waters were mingled together,
> And no field was formed, no marsh was to be seen;
> When of the gods none had yet been called into being.

So begins the Enuma Elish, the seven tablets of creation, describing the ancient Sumerian creation myth.[1] In the beginning the world is without form, and fresh water (Apsû) and salt water (Tiamat) mingle together. Then, in an act of creation, there follow six generations of gods, each associated with a natural manifestation of the world, such as sky or earth. Light and darkness are separated before the creation of luminous objects: the Sun, the Moon, the stars. The sixth-generation god, Marduk, establishes his precedence over all others by killing Tiamat and dividing her body into two parts – the earth below, and the sky above. He establishes law and order – control over the movement of the stars, twelve constellations through which the Sun and the planets move – and he creates humans from mud mixed with the blood of Tiamat.

The similarity to the Hebrew creation mythology described in the book of Genesis has long been recognized:[2] In the biblical story creation takes place in six days, corresponding to the six generations of "phenomenon" gods in Babylon, and the separation of light and dark precedes the creation of heavenly bodies. There is an initial homogeneous state in which the various constituents of the world are mixed evenly together, and an act of creation at a definite point in time – an act which separates these constituents and makes the world habitable (and more interesting).

These aspects are also evident in the Greek creation mythology[3] in which elements of the world are initially mixed together in a formless way – Chaos.

However, at some point, two children are born of Chaos – Night and Erebus, an "unfathomable depth where death dwells" (not an obvious improvement over the initial state of Chaos). But then, also in an unexplained way, something positive and truly magnificent happens: Love is born and Order and Beauty appear. Love creates Light and Day and the stage is set for the creation of the Earth and the starry heavens. Again, separation of light and darkness occurs before the creation of earthly and heavenly objects.

These ancient myths appear to presage the modern scientific narrative of creation – the Big Bang. Here, at a definite point in time, 13.8 billion years ago, the creation of the world occurs. There is no specified prior state and no creating entity; the event is apparently spontaneous.[4] Initially the world is extremely dense and hot but rapidly expands and cools. An instant after the creation event the Universe is recreated in an exponential expansion driven by the energy density of the vacuum; this short-lived inflationary epoch drives the space–time to be very nearly smooth and topologically flat, but at the same time creates tiny fluctuations that eventually will appear as the observed structure of the Universe. A few seconds after creation, when the temperature has fallen to a few tens of billions of degrees, the Universe primarily consists of radiation (photons), electrons and positrons (leptons), ghostly, almost massless neutrinos, a small trace of protons and neutrons (baryons) that will become the primary observable component of the world, and more massive particles that interact very weakly with other components. These massive particles are effectively dark and will forever remain so, although they will come to dominate the dynamics of the matter and the structure that will form from it. After two or three minutes, the light elements (deuterium, helium, lithium) are synthesized via nuclear fusion of about 11% of the protons (plus the remaining neutrons), and the electrons and positrons (anti-electrons) annihilate, leaving a small residue of electrons. After several hundred thousand years, when the Universe has cooled to a few thousand degrees, the protons combine with the electrons, becoming neutral hydrogen, and the radiation (light) decouples from the matter; one could say that light separates from darkness. The structure that is currently observable in the Universe, stars and galaxies and clusters of galaxies, is then free to form via gravitational collapse.

In interesting and perhaps meaningful respects, the form and language of the modern cosmological paradigm is similar to these ancient creation mythologies. In the examples above, there is a time sequence which monotonically and irreversibly progresses from past to present, and there is a definite point in the past at which creation occurs; the world has a beginning. Further, there is initially a more homogeneous state in which the various components are mixed together uniformly; there follows a separation of these components and a separation of light and darkness predating the appearance of astronomical objects; and finally there is the formation of the structures of the present recognizable world. This similarity in the ordering of events – the linguistic framework upon which the creation stories are placed – probably reflects the fact that both the ancient creation myths and the modern cosmological paradigm are products of human thought processes, and limitations of experience and language restrict the description of cosmological phenomena. It would appear as though the language of the scientific paradigm has been filtered through the narrative established by these particularly occidental creation mythologies in ways that are not intentional or conscious. After all, how does one conceive of the spectacular occurrence of the Big Bang, other than through the framework of stories that are in our direct tradition?

But in this connection one should keep in mind that neither in creation mythologies nor in scientific scenarios is it always the case that there is a single universe with a definite beginning. We can identify two trends that run counter to one another: the concept of a temporally finite single world and that of a world that is static eternal and possibly multiple – a universe of universes – co-existing or cyclic. In Hindu creation mythology, the world is created and destroyed and re-created in cycles corresponding to a day and night in the life of Brahma – at 8.6 billion years, amazingly close to the cosmological timescale of the Big Bang.[5] These many sequential universes are rather closely mimicked by current ideas such as eternal inflation and the multiverse scenario, but here the universes are not so much sequential as co-existing (in some extended sense of simultaneity) in different parts of a possibly higher-dimensional infinite stage. But, in the property of its seething, percolating nature, such a world is constant and unchanging. The idea of a universe that is eternally immutable in its general aspect has had great appeal from Newton to Einstein to Hoyle to Linde.

The creation scenarios, ancient and modern, reflect the human drive and desire for an understanding of the Universe, its origin and evolution, and the place of human consciousness within it. However, given the predictive success of modern science and the retreat of superstitious or theistic explanations for the unknown, physical cosmology has achieved a new pre-eminence and dominance in providing a true picture of the Universe – its formation and evolution. With the development of new tools for observing the Universe as a whole, cosmology has become a proper and respectable science. This is particularly true since the discovery of the cosmic microwave background radiation, the CMB – that faint glow of an earlier, hot universe.

In the early twentieth century the development of general relativity, Einstein's theory of gravity, made it possible to "do cosmology" with physics. The new observational tools have now made it possible to "do physics" with cosmology – or, at least, that is the perception. But to what extent do we understand the constituents and evolution of the Universe, and to what extent can we draw conclusions on physics from the cosmology of one universe? Is cosmology a proper physical science?

Certainly in one essential respect cosmology differs from other more traditional sciences: the study of the Universe as a whole lends itself to considerations that are more generally taken to be philosophical rather than purely scientific or narrowly empirical. First of all, there is the issue of reductionism. There is an essentially reductionist predilection in contemporary science in general (a preconception that is not fundamental to the scientific method). Of course, there is a process of unification that is basic to science – a search for efficiency in explanation. But this has been taken quite far: all follows from a few basic laws or a "theory of everything," and to derive such a theory has become a modern holy grail. Even biological evolution and the development of consciousness can be calculated from a few first physical principles (although in practice, this may be impossibly difficult). The reductionist predilection elevates cosmology to a primary role, with large-scale phenomena taking precedence over the phenomenology of individual objects such as stars or galaxies. This prioritization minimizes the possible role and importance of emergent phenomena – the idea that many constructs considered to be basic – even space and time – may emerge, in a way that is not obvious, from more fundamental components. That is to say (with Aristotle) that the total may be more than the sum of its parts.[6]

Emergence may be divided into two categories: "weak" and "strong". Weak emergence is that which can be computed; i.e., the emergent phenomenon can be calculated from the behavior of the primary constituents; weak emergence is not necessarily at odds with reductionism. On the other hand, strong emergence cannot, in principle, be computed and therefore is the more metaphysically challenging concept. But this distinction between strong and weak is in fact blurred by the practical difficulty (or impossibility) of computing complex phenomena from the laws governing the behavior of more basic components. Is it possible to compute the subsequent development of the Universe in all of its present complexity from a snapshot recorded in the cosmic microwave background at the single epoch of decoupling of matter and radiation? In cosmology, the very possibility of emergent phenomena challenges reductionism and the assignment of priority to cosmology over the study of the more complex constituents of the Universe.

A second issue is that of purpose or teleology: modern science is essentially naturalistic and non-teleological, which is to say that evolution, cosmological or biological, has no purpose or final goal but proceeds by natural law without intervention from supernatural or theistic agents. This issue, that of teleology, is a metaphysical (and political) minefield that is best avoided by the practicing scientist. Indeed, it is not possible to be a functioning scientist without being, at least in a methodological sense, a naturalist. But the idea of a purposeless universe would have appeared inconceivable to classical philosophers such as Plato and Aristotle; it is essentially a modern assumption that cannot be proven in a purely scientific context. The possibility of a purposeful (but not necessarily theistic) universe is related to the issue of fine-tuning of fundamental physical constants and cosmological parameters: Why is it that the Universe is conducive to the development of life forms capable of understanding it? A teleological explanation would have profound implications for cosmological and biological evolution, but it is difficult to conceive of how such a concept could be operationally included within the scientific method.

A third issue is that of cosmological space and time. It can be said that the theory of relativity (special and general) destroyed the Newtonian concept of absolute space and time. However, in some sense, physical cosmology has restored it. There is, at least operationally, a preferred universal frame – that which is at rest with respect to the cosmological background – and there is a preferred cosmic time with a preferred direction that apparently proceeds smoothly and irreversibly from the beginning to the present. Time is particularly problematic: Does cosmic time correspond to other measures of time, atomic or biological? Is the direction of time strictly a matter of the entropy of a complex system or does some *a priori* structure (or broken symmetry) set this irreversibility?

Finally, with respect to "doing physics" with cosmology, there is the one-universe problem – the fact that usual statistical arguments cannot be used when confronted with a sample size of one object. Of course, in discussing constituents such as density fluctuations, their formation and evolution, statistical methods are appropriate as long as these fluctuations have a size that is small compared to the limit of the observable Universe – as long as there are many such fluctuations. Then observations of fluctuations and their interpretation belong more properly to the realm of astronomy. However, when the observable fluctuations approach the horizon, the size of the observable Universe, then there are only several in the sample. This is the well-known problem of "cosmic variance," but here again the issue is more fundamental: as in teleology there is a question of "naturalness" or "fine-tuning." Why is the Universe tuned to produce intelligent life?

This question lends itself to an anthropic interpretation. That is to say, the Universe must have quite precisely the properties that it does have (initial conditions, values of fundamental constants, number of dimensions) because these are the very properties that allow cogent observers (us) to develop and pose such questions.[7] There has been much discussion of this idea over the past four decades; supporters argue that it is profound while opponents assert that it is trivial. While the anthropic principle may in some sense be true, it does have the effect of choking off further research into naturalistic explanations for the world being as it is. However, the possibility that the world we see is a selection among many alternatives suggests that there are, in fact, many alternatives. This has led, in part, to the "multiverse" concept – the idea that there are many universes with different initial conditions and possibly different values of physical constants, and we, of course, inhabit one that is conducive to the development of us. Perhaps so, but then any aspect of the Universe that seems unnatural or fine-tuned or improbable can be explained in the context of the multiverse concept. The multiverse is unscientific in the sense that it is not falsifiable. I realize that many sensible people take the idea seriously, but where is its predictive value? Methodologically, the multiverse appears to be as unscientific as the concept of teleology.[8]

These are several of the philosophical issues presented by physical cosmology as a science. But I am not a philosopher and here I will discuss the standard cosmological paradigm in its own terms, both its strong points and what many supporters of the paradigm see as its most puzzling aspect – the result that the two dominant constituents of the Universe, dark energy and dark matter, have an unknown nature and that at the present time there is no independent evidence, other than astronomical, for their existence.

Overall, the current standard cosmological model presents a coherent and consistent picture of observed cosmological phenomena. Much talented effort has gone into interpretation of the now quite detailed observations of the cosmic microwave background radiation and the large-scale observable matter distribution, and it is impressive how well the observations fit together in terms of the standard dark energy–dark matter paradigm. However, these successes have led to a certain degree of triumphalism – the prevalent belief that we now understand the Universe and need only work out the details. It would seem rather dangerous to assume that at this point in human development we possess the tools and the knowledge to explain the entire Universe – its origin and evolution. In one or two generations, such a viewpoint might well appear to be simplistic and naive. So here I plead for a bit less hubris. After all, hubris angers the gods and pride is a deadly sin.

2
Three Predictions of Physical Cosmology

2.1 The Basis of Physical Cosmology

The scientific method requires the use of empirical evidence in discerning the nature of the world; one does not rely on folklore, prejudice, tradition, superstition, theology or authority but only upon the evidence that is presented to the senses, most often enhanced by the appropriate instrumentation. Significantly, science is not only observation and classification. The scientist seeks to unify diverse phenomena by a set of rules or theories that are rational, mathematical and general (although perhaps not always obvious or belonging to the realm of common sense). Ideally, theories should make predictions that are testable by repeated experimentation or observation. In this way the theory can be supported or rejected. The underlying assumption is that the world is rational and understandable to human beings, an assumption that appears to be justified by several hundred years of experience.[1]

In physics, the first outstanding application of the scientific method was that of Isaac Newton. Building on the empirical descriptions by Galileo in codifying the kinematics of objects locally, and by Kepler in his formulation of the laws of planetary motion, Newton was able to unify the phenomenology of falling objects on the Earth with that of the planets moving about the Sun by a single law of gravitation combined with basic laws of motion. Newtonian dynamics provides the prototypical example of the scientific method in the steps followed to derive a general theory. The most dramatic early success of Newtonian theory (prediction as opposed to explanation) was its application by Urbain Le Verrier in the mid nineteenth century to predict the existence and position of the planet Neptune, using the observed small deviations in the motion of Uranus from that calculated due to the known massive bodies in the Solar System.

Such predictions form the core of the scientific method, and here I consider the three essential predictions of modern – twentieth century – physical cosmology: the expansion of the Universe; the existence of the cosmic background electromagnetic radiation; and the presence of small angular fluctuations in the intensity of that radiation. These are true predictions. They arose from theoretical considerations before being confirmed by observations, and as such, they form the backbone of physical cosmology.

Physical cosmology is the application of the scientific method and the laws of physics, in so far as these are understood, to the study of the Universe as a whole. It is assumed that the laws determined locally apply everywhere at all times. The origins of physical cosmology – the modern study of the structure and evolution of the Universe – can also be traced to Newton. It was imagined by Newton and his contemporaries that the Universe was an infinite, homogeneous and static arrangement of stars with the mass of a spherical shell about any point increasing as the square of the radius of the shell. The question naturally arose of how such a system would evolve given the inverse square gravitational attraction between these constituents of the world. That question was specifically put to Newton in correspondence with a theologian, Richard Bentley, in 1692.[2] Bentley was worried about the infinitude of the Universe and the resulting infinity of gravitational force on any one particular object. Newton, of course, realized that if all such objects were symmetrically placed then the force from all sides would perfectly balance – a sort of balance of infinities. However, he also realized how unlikely such a perfect arrangement of masses would be: "a supposition fully as hard as to make the sharpest needle stand upright on its point upon a looking glass" – in fact, he added, the balancing act would be required of "an infinite number" of needles. Newton understood the inherent instability of this arrangement: if one star were displaced very slightly, it would begin to move, disrupting the entire arrangement and leading to collapse.

This issue was dealt with in some mathematical detail by Hugo von Seeliger in 1895, who proved that the gravitational potential of a particle in a homogeneous universe was infinite and the force was indeterminate. This prompted von Seeliger to suggest a modification of Newton's law by adding an exponential cutoff beyond a certain length scale; that is to say, the gravitational attraction effectively vanishes beyond a certain distance, thus avoiding the problem of infinities.[3]

A different solution of this problem is to suppose that the Universe is inhomogeneous – that it consists of a great disk of stars suggested by the appearance of the Milky Way, as mapped in detail by William Herschel (1781) and later by Jacobus Kapteyn (1900). The stars are moving in such a way as to counteract the force of gravity and thus maintain the overall shape of the configuration. However, in the eighteenth century, Immanuel Kant had, in effect, questioned this approach by his proposal that the Universe was filled with "islands" of disks and that the Milky Way was only one such island; the implication was that on a large scale the Universe returns to homogeneity.

One hundred years ago, Albert Einstein published his famous General Theory of Relativity, a theory of gravity in which the single gravitational field of Newton is replaced by the geometry of a four-dimensional space–time; in effect, Einstein "geometrized" gravity. The topology of this space–time is related to the matter–energy distribution by field equations, analogous to that of Newton which relates the gravitational field to the density distribution of matter. Particles follow special paths in the space–time that has been molded by the matter–energy distribution – paths called "geodesics" that are essentially the shortest distance between two points or events in four dimensions (relativistic particles such as photons follow special geodesics – "null" geodesics – in which the separation between events is zero). As summed up succinctly by John Wheeler many years later, "matter tells space–time how to curve; space–time tells matter how to move."[4]

In the light of his theory, Einstein reconsidered the problem of a homogeneous, *static* universe in 1917. In the spirit of von Seeliger's modification of Newtonian gravity, Einstein added a constant term to his field equation – the cosmological constant – a term that, in the Newtonian limit, provides a universal repulsive force that increases linearly with distance from any given point and can therefore perfectly balance the attractive gravity force within a homogeneous medium – that is, *if the medium has a preferred density*.[5] There is the rub. If the density varies slightly from this preferred value, the Universe will expand forever or contract to a point within a finite time; the configuration appears to be unstable, or at least a static solution is very special indeed. Willem de Sitter quickly pointed out that a universe with a positive cosmological constant, but devoid of matter or energy, would be in a steady state of constant exponential expansion, and his model, relevant to the current cosmological model, has become known as the "de Sitter universe."

2.2 The Expansion of the Universe

Of course the world does contain matter, so what does general relativity say about the "real" world? This problem was considered by a young Russian physicist–mathematician, Alexander Friedmann (1924), who provided a solution of Einstein's field equation (unmodified by a cosmological constant) for an infinite homogeneous medium. The answer is simple: the Universe is not static but dynamic. General relativity (without Einstein's cosmological constant) permits only solutions that are uniformly expanding or contracting. In fact, there are three such general solutions (see Figure 2.1). The first is a bound universe that would

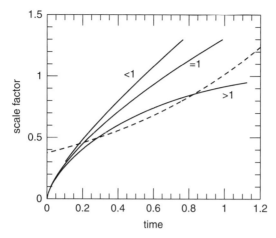

Figure 2.1. The solid curves show the form of the expansion for the three general classes of Friedmann models. The horizontal axis is the cosmic time in units of the Hubble timescale (14 billion years by present estimates) and the vertical axis is the universal scale factor: a dimensionless number by which all separations are multiplied. The curve labeled "> 1" is for the closed universe with a density larger than the critical density; "= 1" is the flat universe with a density equal to the critical density; and "< 1" is the open universe with a less than critical density. The dashed curve shows the form of the expansion for a de Sitter universe dominated by a cosmological constant.

expand to a certain point and then re-collapse; the second is a model in which the expansion, after an infinite time, slows to zero; and the third is an unbound universe in which the expansion decelerates but continues forever. Which model universe we live in depends upon the average density of energy–matter at the present epoch; there is a critical value of the density for which the Universe is in the marginal state between eventual re-collapse and eternal expansion.[6]

The overall topology of the Universe also depends upon this critical density. In the case of a re-collapsing universe (the density is higher than critical), the space–time is positively curved, as a sphere; i.e., the Universe is finite but without boundary. In the eternally expanding universe (less than critical density), space–time is negatively curved, like a saddle (although it is a strange saddle – any observer is always found at the midpoint of the saddle seat). Of course, such a universe is infinite in extent. In the marginal case in between (precisely the critical density), space–time is flat and also infinite. The significant point is that, in all cases, *the universe must be dynamic, not static*, and this was the first essential prediction of physical cosmology.

The reality of a non-static universe was confirmed by observations that caught up with theory several years later in a remarkable series of astronomical discoveries, First, Harlow Shapley (1914), exploiting the discovery of the period–luminosity relation of Cepheid variable stars by Henrietta Leavitt (as well as the new 100-inch telescope on Mt. Wilson in California), observed these variable stars in the halo of globular clusters surrounding the Milky Way and so determined the scale and the position of the Sun within this great stellar system. The Milky Way, it turns out, is an enormous disk of stars, tens of thousands of light years across, with the Sun occupying an obscure position in the outskirts; in locating the Sun near the center of the system Herschel and Kapteyn had been fooled by the obscuration of starlight by interstellar dust in the plane of the disk. Then Edwin Hubble (1922), also using the 100-inch telescope, detected such Cepheid variable stars in the Andromeda spiral "nebula" and applied the period–luminosity relation to establish that the distance is, incredibly it seemed at that time, more than one million light years – Andromeda is not a satellite of the Milky Way as Shapley thought, but an enormous stellar system with a scale comparable to that of the Milky Way. The same must be true of the other spiral nebulae distributed more or less isotropically around the sky; the island-universe picture of Kant is essentially correct.

Vesto Slipher reported spectroscopic observations (1915–1918) of spiral galaxies, then called spiral nebulae; he found that the common spectral lines arising from elements such as hydrogen and calcium are not at the wavelengths measured in the laboratory, but are shifted most often toward the red end of the spectrum; i.e., they are typically redshifted. The shift in spectral lines, usually designated by z, is given by the change in wavelength, $\Delta\lambda$, divided by the rest wavelength, λ_0; i.e., $z = \Delta\lambda/\lambda$ (by convention positive z is motion away from us, negative is toward us). If interpreted in terms of the Doppler effect, the velocity of the emitting object toward or away from the observer would be $v/c = z$ for z much less than one. So the implication is that the spiral nebulae are moving away from the Milky Way. Knut Lundmark (1924) estimated the distances to these galaxies with measured Doppler shifts. He did this by assuming that the galaxies were standard candles (with a characteristic luminosity) and standard meter sticks (with a characteristic size). He found that the velocity of recession was proportional to the distance. This was repeated by Edwin Hubble in 1929, who made his own distance estimates using Cepheid variables, and found indeed that these relatively nearby galaxies appeared to be rushing away from the Milky Way with a speed proportional to distance (Hubble eventually received full credit for the discovery, but that is another story).[7] The original Hubble diagram is shown in Figure 2.2; apart from a problem of distance calibration this reveals the predicted expansion of the Universe.[8]

These results imply that the Universe is truly dynamic and uniformly expanding; the relevant parameter is the constant of proportionality between velocity and distance, the Hubble constant with units of inverse time, H, as it has come to be designated.[9]

2.2 The Expansion of the Universe

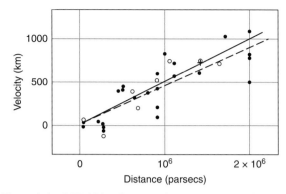

Figure 2.2. The original Hubble diagram for nearby galaxies as presented by Hubble in 1929. The horizontal axis is the distance in parsecs and the vertical axis is the recession velocity as measured by the Doppler shift of spectral lines. Hubble's calibration of distance indicators (Cepheid variables) was systematically in error so the Hubble constant (≈ 500 km/s per Mpc) was much too large, but the proof of a non-static universe is clear.[8]

It was not fully appreciated at the time, but this observation of *the redshift–distance relation was the first predictive success of physical cosmology*. Slipher, Hubble and Lundmark were not looking for a predicted effect; they were conducting a systematic program of astronomical observations to measure spectra and determine the distances to spiral galaxies. Initially Hubble himself was not fully convinced that redshift implied recession. Among the theoreticians Arthur Eddington, Willem de Sitter and Einstein, there were discussions about the cosmological significance of this observation. In the 1920s Friedmann's work in turbulent, revolutionary Petrograd was not generally known, and the observational result was more usually compared to de Sitter's universe dominated by Einstein's cosmological constant. But gradually, in the 1930s, Hubble's result came to be interpreted, by most astronomers and physicists, as indicating the true expansion of the Universe as described by Friedmann.

If the Universe is getting bigger, then it must have been smaller in the past. Extrapolating the expansion backward, at some point in the past all of the galaxies must have been in contact – the matter of the Universe was much denser than at present; perhaps the Universe had an explosive origin and a finite age comparable to the inverse of the Hubble constant, $1/H$. This thought had occurred to a Belgian priest, Georges Lemaître, in 1927, two years before Hubble's famous paper. Independently of Friedmann's calculation, Lemaître also recognized that Einstein's theory required a dynamic universe and he proposed an explosive expansion, in the first version of what would become known as the "Big Bang."

Lemaître thought in terms of a primordial atom: In the beginning all the matter in the Universe is in a single atom of enormous atomic mass. This atom is of course unstable to nuclear fission and splits instantly, spraying smaller atoms into all directions, creating the expanding Universe and the abundances of chemical elements.

Arthur Walker and Howard Robertson independently considered the problem of an isotropic homogeneous universe; that is to say, a universe that appears the same to all observers anywhere, independent of the direction they look, and one in which the average physical properties are the same everywhere at a given cosmic time. For such a universe, they were able prove that there is one and only one general form for a space–time that satisfied these conditions; that is, there is a unique form for the gravitational field in Einstein's theory (the space–time metric).[10] This result is independent of Einstein's theory, but it must be the exact solution of the field equations of general relativity in an isotropic homogeneous universe. The kinematic results of observational cosmology can be derived using only this form of the space–time, but then applying the Einstein field equations to this space–time leads directly to the Friedmann models. Thus the standard general relativistic framework of the universe dominated by matter–energy has become known as the Friedmann–Lemaître–Robertson–Walker, or FLRW models. These are distinguished from the de Sitter universe, which is empty of matter or radiation, but with exponential expansion caused by a positive cosmological constant.

2.3 The Background Radiation

In the 1930s and 1940s, concepts of modern physics, such as the Planck radiation law, began to be increasingly applied to cosmology. At the beginning of the twentieth century, Max Planck had ushered in the quantum revolution by proposing that electromagnetic radiation could exist only in discrete packets – multiples of a basic unit given by $h\nu$, where h is a new constant of nature and ν is the frequency of the radiation. By this proposal, Planck was able to explain the run of intensity with wavelength for thermal radiation – radiation from a hot object, the so-called *black body radiation*. These quantized packets of energy are now called *photons* and it is recognized (largely due to the work of Einstein) that they have many of the properties of particles.

If the Universe contains electromagnetic radiation – that is, a gas of these photons – then in the early expansion when the radiation was scattered and absorbed repeatedly by the ionized matter, the photons would take the form of thermal radiation having the characteristic Planckian black body spectrum with a characteristic temperature. Moreover, as the Universe expands, the radiation maintains this form but the temperature decreases as the scale of the Universe becomes larger; the radiation redshifts due to the expansion of the Universe. It follows then that the energy density in the radiation decreases more rapidly with expanding volume than does the energy density in the matter (not only does the density of photons decrease inversely with the expanding volume, but the energy per photon decreases due to the redshift). However, going back in time, this means that the radiation would dominate the energy density of the Universe at sufficiently early epochs – going back into the past, at an earlier epoch the Universe becomes a radiation universe and not a matter universe.[11]

A second concept of increasing significance for astrophysics and cosmology was that of nuclear reactions and transmutation of chemical elements. By 1940 it was generally recognized that the energy source of stars – the mechanism that permits the stars to shine for billions of years – is nuclear fusion. As described independently by Hans Bethe and Carl Friedrich von Weizsäcker, at the temperatures and densities prevailing in stellar interiors four hydrogen nuclei – protons – can combine to form a single helium nucleus by two different processes.[12] The mass of a single helium nucleus is about 0.7% less than that of four protons, the missing mass appearing as energy via Einstein's famous formula. This process could work in principle up to iron but beyond that, fusion would require energy, not yield energy. But both Bethe and von Weizsäcker realized that at the temperatures and densities prevailing in the interiors of normal stars like the Sun, elements heavier than helium could not be formed, and concluded that the formation of heavier elements "took place before the formation of the stars."

This possibility led von Weizsäcker to propose his own Big Bang model; he suggested that a primordial cloud of hydrogen collapsed under its own gravity and so converted its entire nuclear binding energy into heat, causing the initial explosion – a Big Bounce. Within this primordial gas cloud the chemical elements were synthesized in an equilibrium process; i.e., the ratios of the abundances of elements with nearby binding energies were given by the Boltzmann equation.[13] The model was in many respects similar to Lemaître's primordial atom, and although Weizsäcker did not include the concept of relativistic expansion à la Friedmann, he did present a coherent proposal for the origin of the Big Bang and the chemical elements.

This idea sparked the interest of George Gamow, who realized that such an equilibrium origin of the elements would not be possible in the context of the Friedmann equations; the time for the Universe to expand significantly was shorter than the timescale for nuclear reactions to establish equilibrium. Gamow was a Soviet nuclear physicist who had attended lectures by Friedmann in Leningrad. In 1934 he emigrated to the West and finally to the United States, where he began to contemplate the energy sources of stars and the origin of the elements. He accepted the result of Bethe and von Weizsäcker that the heavy elements could not be synthesized in stars, but thought that the elements might be formed in the early expanding Universe via the process of neutron capture onto protons and other atomic nuclei. Free neutrons (not bound in an atomic nucleus) have a lifetime of about ten minutes, so this would have had to happen in the first minutes of the Big Bang; following Gamow, synthesis of the elements would not require a primordial atom or collapsing gas cloud, but could occur in the dynamic expanding universe of Friedmann and Lemaître.

In 1948 this idea was developed further as a radiation-dominated universe – the hot Big Bang – in a famous paper by Gamow, in collaboration with his student Ralph Alpher and colleague Hans Bethe, and in later work by Alpher and Robert Herman. Even though it was subsequently shown that elements heaver than helium could not be produced by this mechanism of neutron capture (due to the mass-four barrier – the fact that there is no stable mass-five element) this work led to the second major prediction of physical cosmology: *the cosmic microwave background radiation*. To produce the light elements in the hot expanding universe, the temperature, which in the radiation-dominated universe is directly related to the density of protons, had to be several tens of billions of degrees within a few minutes of the creation event. In the subsequent expansion of the Universe, this primordial radiation would have cooled to about five degrees above absolute zero at the present epoch (as estimated by Gamow and collaborators); in other words, the Universe should have been flooded by black-body radiation, with its peak intensity at a wavelength of approximately one millimeter – microwave radiation.[14]

At the time (the early 1950s), not much was made of this prediction because the detection of such universal radiation was beyond technical capabilities. Also, there was a competing cosmological theory, an alternative to the Big Bang, which had some following at the time: the Steady State universe, developed by Hermann Bondi, Thomas Gold and Fred Hoyle. This model was based upon the so-called "perfect cosmological principle" – the idea that the Universe not only appears to be isotropic and homogeneous to observers at all points in space, but also to all observers at all points in time; the aspect of the Universe in its average properties does not change with time. There is no beginning, there is no end. The constant Hubble parameter implies that the expansion law is exponential, as that of the de Sitter model.[15]

The prospect of an eternal unchanging universe has a certain philosophical appeal, but would not the observed expansion of the Universe imply that the average density should decrease? The solution proposed by Bondi, Gold and Hoyle was that creation is a continuous process – that new matter is continually created to fill the expanding vacuum and that new galaxies are continuously being formed to maintain a constant average density of galaxies. The basic ingredient is hydrogen, with heavier elements continuously produced in the interiors of stars before being spewed out into the interstellar medium by various mass-loss processes: stellar winds, planetary nebulae events, novae and supernovae. This, of course, is contrary to the original conclusion of Bethe and von Weizsäcker that heavy elements could not be synthesized in normal stars, but that was before the full range of conditions in stellar interiors was appreciated.

In the mid 1950s, with the advent of large computers, the theory of stellar structure and evolution took a great leap forward, thanks largely to the efforts of Martin Schwarzschild. Schwarzschild demonstrated that normal stars like the Sun evolve by depleting the hydrogen in the very central regions, thus developing a core of helium. As the helium core builds up, the star expands, cools and becomes more luminous; it becomes a red giant. The core is extremely hot – 100 million degrees, as opposed to the 10 million degrees of the interiors of solar-like stars – and extremely dense – like a white dwarf with a density at the center of one million tons per cubic meter. Under these conditions, nucleosynthesis can leap around the mass-four barrier; three helium nuclei can fuse to form carbon (with the intermediate step of forming beryllium) and the way is open for the formation of iron. Thus a pre-stellar origin is not required for the chemical elements beyond helium.[16]

The theory of stellar evolution also appeared to present a serious problem for the Big Bang model. An evolving star traces a definite path in a diagram of luminosity (power) versus temperature (the Hertzsprung–Russell diagram). More massive stars evolve more quickly along this path, so if there exists a collection of stars over a range of masses that all form at the same time, the observed temperature–luminosity plot of this ensemble of stars would reveal this characteristic pattern of evolution. There are such groups of stars that are thought to be coeval – the globular clusters found in a large halo about the disk of the Milky Way. By comparing the theoretical luminosity–temperature relations of evolving stars of differing masses to those observed for individual globular clusters, it became possible to estimate the age of globular clusters. In several cases, these turned out to be larger than the age of the Universe as estimated by the observed Hubble constant ($1/H_0$); it is clearly impossible for a constituent of the Universe to be older than the Universe itself. This would not be a problem if the Universe were infinitely old as in the Steady State model; then the inverse Hubble constant would have no significance as an age (as it turned out, this was not a problem at all because of early overestimates of the Hubble constant).

In 1957 there appeared the famous work on stellar nucleosynthesis, combining theory and observations – a seminal paper authored by Margaret Burbidge, Geoffrey Burbidge and William Fowler, in collaboration with Hoyle.[17] They demonstrated that not only the elements lighter than iron, but also the heavier elements in their observed abundances, can be produced in the interiors of stars by neutron capture onto iron nuclei. A complete and successful alternative to cosmic nucleosynthesis had been presented.

The Steady State seemed to be viable. There was, however, one serious problem pointed out by Hoyle himself in collaboration with Roger Taylor.[18] The abundance of helium was known to be about 8% by number of atoms. To produce this number of helium nuclei by the fusion of hydrogen in the interior of stars would release ten times more stellar radiation than is observed in the Universe. Where is this radiation?

In 1965, Arno Penzias and Robert Wilson discovered the cosmic microwave background, the CMB, an unavoidable prediction of a Big Bang cosmology hot enough to produce the primordial abundance of helium. Penzias and Wilson were not looking for the background radiation; the discovery was serendipitous. These two physicists at Bell Telephone Labs in Holmdel, NJ, were trying to build a very sensitive radio telescope – a radio receiver with very low noise – but they found that no matter which way they pointed their horn-receiver, there was a persistent static corresponding to a temperature of about three kelvin.

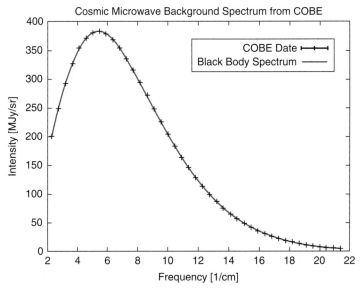

Figure 2.3. The spectrum of the cosmic microwave background as measured by the COBE (Cosmic Background Explorer) satellite compared to the theoretical spectrum of a black body with a temperature of 2.783 kelvin. The horizontal axis is the inverse wavelength (1/cm) and the vertical axis is the intensity of the radiation. The points are indistinguishable from the curve, making this the most perfect black body astronomically measured.[20]

Down the road at Princeton University, there was a group of physicists headed by Robert Dicke that actually was searching for the cosmic background radiation. Dicke was unaware of Gamow's earlier work, but had developed his own theory of an oscillating universe which went through hot phases. He predicted the presence of a background radiation with a temperature of ten kelvin, a relic of an earlier hot phase, and his group was constructing radio receivers to detect it. When they heard of the Penzias and Wilson observations, they realized immediately that *the primordial black body radiation – the second prediction of physical cosmology – had been found*. This observation put to rest, at least for most astronomers and physicists, the Steady State cosmology, although several of the proponents of the picture fought a rear-guard action for decades.[19] In 1994 the spectrum of the CMB was observed with great precision by the COBE satellite; this is shown in Figure 2.3. The Planck spectrum with a temperature of 2.738 degrees matches the observations perfectly, leaving no doubt of the black-body origin of the radiation.[20]

2.4 Anisotropies in the Background Radiation

The third prediction of physical cosmology came within two years of the discovery of the background radiation, but it concerned a problem that had been around since Newton and Reverend Bentley – the gravitational instability of a homogeneous medium and the origin of structure in the Universe. Physical cosmology is built upon the Cosmological Principle – the assumption of isotropy and homogeneity of the Universe on the large scale. This assumed symmetry greatly facilitates the solution of the field equations of general relativity and the calculation of the Friedmann models. But of course, it is evident from observations that the present Universe is not homogeneous at all on scales smaller than its characteristic size, the Hubble radius (c/H): there are stars and galaxies and clusters of galaxies and clusters of clusters of galaxies. There is a richness of structure in the Universe on scales ranging up to many millions of light years. What is the origin of this structure in the context of the Big Bang model?

The first modern mathematical treatment of this problem was by Sir James Jeans in the beginning of the twentieth century. Jeans considered the stability (or instability) of an infinite static homogeneous gas with a finite density and temperature. He mathematically followed the development of small density fluctuations in this medium by ignoring the original Newtonian problem of an infinite gravitational potential in the unperturbed homogeneous medium. Jeans found that the fluctuations were stable (no collapse) if they were smaller than a certain critical length scale that depended upon the temperature and density of the medium ($\propto \sqrt{T/\rho}$). Below this length scale, ordinary gas pressure is sufficient to prevent the collapse of any over-dense region. But if the fluctuations are larger than this scale, the *Jeans length*, they collapse; the density of large fluctuations grow exponentially on a timescale inversely proportional to the square root of the density.[21]

Jeans' analysis applied to a static medium; however, the Universe is not static but uniformly expanding. How do fluctuations develop in such an expanding medium? This problem was solved in the context of general relativity in 1948 by Evgeny Lifshitz, who demonstrated that the Jeans criterion for collapse still applied: fluctuations larger than the critical length scale collapsed, but it took longer than in a static medium. In an expanding medium, the relative density of unstable fluctuations grows as a power law in time and not as an exponential.

Gravitational instability became the favored mechanism for forming structure in the Universe. The Universe at early epochs is really very homogeneous and isotropic, but the presence of small fluctuations allows structure to form to the extent that we now observe.

After 1965 we knew, however, that there is cosmic radiation in the mix. How do the matter fluctuations necessary for the formation of structure affect this radiation? This was a problem first considered in a very precocious work by Rainer Sachs and Arthur Wolfe in 1967.[22] They demonstrated that the presence of density fluctuations leaves an imprint on the radiation content of the Universe. Basically, density fluctuations of a certain magnitude, $\Delta\rho$, measured in terms of the average density ρ, cause comparable fluctuations in the intensity of the thermal background radiation measured in terms of temperature: $\Delta T/T \approx \Delta\rho/\rho$.

This relationship results from two competing effects. The first is adiabatic compression: when a gas is compressed (a positive density fluctuation) the temperature increases; photons from over-dense regions appear hotter – bluer. The second mechanism, however, goes the other way. Photons climbing out of the gravitational potential wells caused by the positive density fluctuations are gravitationally redshifted; this makes the photons appear cooler – redder. The second effect is a bit larger, so there is a net fluctuation of the photon temperature corresponding to positive and negative density fluctuations (cooler and warmer). Given the observed structure in the Universe, but the homogeneity of the Universe at early epochs, these fluctuations in the cosmic background radiation must be present and observable at some level. *The existence of large- and small-scale fluctuations in the CMB is the third prediction of physical cosmology.*

At the time of the prediction, however, in spite of observations of increasing sensitivity, the CMB appeared to be extremely smooth and isotropic – certainly down to levels of a few percent. The first anisotropy in the CMB became apparent

2.4 Anisotropies in the Background Radiation

in the early 1970s and was observed with high significance in 1976. Using a detector flown in a high-altitude aircraft, a group headed by George Smoot from the University of California at Berkeley found a large-scale variation at the level of 1 part in 1000. It is a "dipole" variation – one half of the sky is very slightly warmer than average and the other half is cooler – and is certainly due to the "peculiar" motion of the Earth with respect to the universal frame defined by the CMB; in a sense, the aether drift, looked for by Michelson and Morley a century before, had been found. The local group of galaxies that includes the Milky Way appears to be moving at over 600 km/s with respect to this universal frame. While this is of considerable interest, it is not directly related to the density fluctuations which form the wealth of structure in the Universe.

As the observations pushed the limits down by another factor of 10 (less than one part in 10 000), the continued null results became disturbing. The microwave photons that we observe now last interacted with matter when the temperature was about 3000 K (at a redshift of about 1100). At earlier epochs, the hydrogen and helium atoms of the Universe are fully ionized and the radiation (the photons) is coupled to this ionized gas via interaction with the free electrons (electron scattering). In this state no structure can form because the Jeans length exceeds the size of the causally connected region, the horizon – the distance that light can travel during the lifetime of the Universe at a particular epoch. Essentially, the pressure force provided by the photons is too high to permit gravitational collapse of a causally connected region. But when the temperature cools below 3000 degrees, the protons combine with the free electrons to form neutral hydrogen and the photons are free to stream directly to us at the present time. So in a real sense, the CMB photons that we currently observe were emitted by an opaque wall of 3000 degrees, and the predicted fluctuations in the background temperature reflect the density fluctuations in the gas at that time (when the Universe was about 380 000 years old). But to form observed structure, the density fluctuations must grow by a factor of at least as large as the redshift, 1000, which implies that they should be present at the earlier epoch at the level of 1 in 1000. Where are they?

This issue was resolved following the launch of the COBE satellite (the "Cosmic Background Explorer") in 1989. After two years of gathering data, temperature fluctuations on the angular scale of tens of degrees were detected in the CMB at the level of several parts in 100 000. *The mechanism creating the small intensity variations in the background radiation was exactly that described by Sachs and Wolfe twenty-five years earlier; the third prediction of the Big Bang was verified –* with one caveat. These angular scales corresponded to structures which would now be on a linear scale of two or three billion light years – much larger than the scale of observed structure. The observed fluctuations could not possibly have led to the

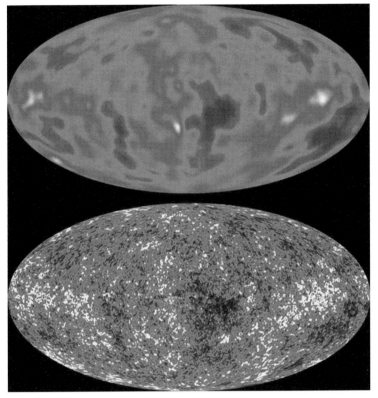

Figure 2.4. The top panel shows an all-sky map of the anisotropies in the microwave background as observed by the COBE satellite on an angular scale of tens of degrees. The lower panel is the same but with the much higher resolution of ten minutes of arc provided by the *WMAP* satellite. The fluctuations seen with COBE are much too large to form the currently observed structure in the Universe – they would not yet have collapsed – but the observation did verify the predicted fluctuations that can lead to structure via gravitational collapse. Credit: NASA-WMAP science team.

objects we actually observe in the Universe at present; in fact they would not have had sufficient time to collapse to form structure.

Detection of fluctuations on smaller scales would have to wait for another several years, but COBE demonstrated that the predicted fluctuations were in fact present – a convincing justification of the Big Bang scenario. The COBE observations are shown on an all-sky map in Figure 2.4; higher-resolution observations made with the *WMAP* satellite ten years later are shown in the lower panel.[23]

To summarize, the three fundamental predictions of physical cosmology – the Big Bang model – were: (1) a dynamic, or expanding universe; (2) the presence of a cosmic background radiation with a temperature of a few degrees above absolute zero; and (3) the presence of small variations in that temperature that reflect the density fluctuations that lead, via gravitational collapse, to the observed structure in the Universe. These are true predictions – they were proposed, although not fully appreciated, before being verified by observations – a spectacular success of physical cosmology and a justification of the assumptions underlying the scientific method.

It is now difficult to imagine that this overall picture of the world could be fundamentally wrong. The Universe (our Universe, at least) really does have a beginning and has evolved over time. At earlier epochs the aspect of the world was very different than at present: it was much denser, hotter and more homogeneous (smoother) than we observe it to be at present. In this early state the medium contained tiny density fluctuations that would grow by gravitational collapse to become the wealth of structure that we observe at present. If we accept at all the premises of the scientific method, the modern creation scenario certainly provides a truer picture of reality than do the ancient creation myths. However, this does not mean that the structure and evolution of the Universe are completely understood and that there are no limits on cosmology as a science. There is a problem of initial conditions. Why is the Universe as revealed by the CMB so smooth and isotropic? What is the origin of the small density fluctuations that become the observed structure? Did the Universe emerge from a singularity where general relativity, and hence the Friedmann equations, are invalid? We now consider an attempt to provide a naturalistic answer to these questions – a proposed event that has become an essential aspect of the cosmological paradigm: inflation.

3

The Very Early Universe: Inflation

3.1 Fine-Tuning Dilemmas and the Initial Singularity

The basic scenario of the Big Bang was in place by 1970. It was generally accepted that the Universe was once much hotter and denser and more homogeneous than at present. There was an origin of the world at a definite point in the past more than 10 billion years ago, and the light chemical elements, primarily helium, were synthesized in the hot early history of the cosmos. These were the essential aspects of what was becoming the standard scientific cosmological model. There were, however, several problems in principle with the Big Bang – issues that were not so much phenomenological as aesthetic – basically problems of naturalness.

In 1970 Robert Dicke emphasized that there are fine-tuning problems in the FLRW models for the Universe.[1] Why is it, for example, that the density in the present Universe is so nearly equal to H_0^2/G, where H_0 is the present value of the Hubble parameter and G is the constant of gravity (H varies with time – it is not a true constant)? In other words, why should the present density be so close to the critical density for a flat universe if it is not, in fact, precisely the critical density?

This question arises because of the nature of the Friedmann equations. The measure of the average density of the Universe is typically given in terms of the critical density and is designated as Ω. This parameter also varies with cosmic time, but if Ω is greater than, less than or equal to one, it is always so (the curvature of the Universe cannot change with time). The value of the density parameter at the present epoch is Ω_0 (analogous to the Hubble parameter H_0), so $\Omega_0 = 1$ (implying a density of about 10^{-29} grams per cubic centimeter at present) exactly corresponds to the flat FLRW universe that asymptotically expands forever. The problem is that for Ω_0 to be close to one but not one requires incredible fine-tuning of the density in the early Universe. If, for example, $\Omega_0 = 0.05$, then at the time of nucleosynthesis, when the temperature of the black body radiation is on the order of 10 billion degrees, the density parameter of the Universe could not deviate from one to within several parts in ten billion; for earlier epochs (higher energies) this tuning problem becomes more severe. The point is that $\Omega = 1$ is an unstable equilibrium point for the Friedmann equations; a very slight deviation from one in the beginning rapidly diverges from one at later times – so, more naturally, a Friedmann universe should be essentially empty or it should have collapsed within a few nanoseconds of the creation event. This appears to be a problem of extreme fine-tuning of the initial density of the Universe.

There is a second problem of naturalness that may be put in the form of a question: Why is the Cosmological Principle – basically the assumption that the Universe looks the same in all directions (isotropy) – so valid as a working assumption? When we observe the CMB, we are looking at the opaque wall at a redshift of about 1100, when the temperature of the black body radiation was 3000 degrees and the primordial photons last interacted with matter. The Universe was about 380 000 years old at this point, which means that light waves could travel 380 000 light years. This roughly defines the horizon – the size of a causally connected region – that obviously increases with cosmic time (currently the horizon would be on the order of 14 billion light years). At the time of decoupling, a horizon would from our point of view intersect a little over one degree on the sky, which means that pieces of this wall with an angular separation of more than one or two degrees would not be in causal contact with each other at that time. Moreover, in the context of the FLRW models, they could never have been in causal contact. So how could patches of the sky that were not previously in physical contact have a temperature so very nearly the same? What tells a patch of sky that it should have the same temperature as another patch on the opposite side of the sky? (This is more commonly called the "horizon problem.")

There was, however, another problem in principle with the Big Bang – a problem not of fine-tuning but of great concern to theoretical physicists of the 1960s. In the context of the Big Bang, going back in time the Universe becomes more dense and hotter. That means that the energies of photons and particles are higher at earlier epochs. At the epoch of nucleosynthesis, when the Universe is 100 seconds old, the particle energies are on the order of 100 000 electron volts (100 KeV), but when the Universe is only 0.03 seconds old, the particles have energies of 1000 million electron volts (1000 MeV). This is on the order of the rest mass energies of protons, so protons and anti-protons are broken down into their constituents – quarks. There is a sea of quarks and high-energy photons in roughly comparable numbers. These particles are highly relativistic because their total energy exceeds the rest mass energy, so, in this sense, the kinetic energy of particles is synonymous with mass. The difference between particles with non-zero rest mass and photons is blurred in these early times. Proceeding further back in time, the energy of the particles becomes more and more extreme.

How far back can the Big Bang be extrapolated? How far up can the energy of particles and photons be pushed in this backward evolution? How close to the creation event might we expect the picture of FLRW expansion to be relevant? Because Friedmann expansion is based upon general relativity, the question then becomes one of the energy or cosmic age at which general relativity breaks down. At some early time, quantum effects, perhaps the discreteness of space and time, would come into play and a theory of quantum gravity would be required. It is natural to suppose that this transition occurs near the Planck energy – the unit of energy that can be constructed out of the basic constants of quantum mechanics and gravity: Planck's constant $\hbar$ (actually Planck's constant divided by 2π), the constant of gravity G and the speed of light c – that is $\sqrt{\hbar c^5/G}$, about 10^{28} electron volts or 10^{19} GeV (giga-electron volts). This energy would correspond to a five-kilogram bowling ball being hurled at 100 000 km/hour – quite an energy for a single photon or sub-atomic particle. The age of the Universe when particles achieve the Planck energy would be comparable to the Planck time $\sqrt{\hbar G/c^5}$, which is 10^{-44} seconds. We would not expect the standard Friedmann cosmology to be valid at an age earlier than the Planck time; to describe earlier epochs we would require a more complete theory of physics at these extremely high energies.

But here is the problem in principle: at this early age, particles and photons are so energetic and therefore have such a high mass that they in effect become black holes; the Compton radius, that measure of the quantum spreading of a particle, becomes comparable to the Schwarzschild radius. This means that particles and/or photons are contained within their horizon which, as we know from the time of Karl Schwarzschild and Arthur Eddington, hides a singularity. So an extrapolation of FLRW back to the Planck time suggests that the Universe was born out of a singularity. As supposed originally by Richard Tolman, it is likely that new physics, part of a theory of quantum gravity, intervenes at or near the Planck scale; this speculative possibility may provide one escape from the singularity trap. But there is another possible exit.

3.2 An Early De Sitter Phase

In the mid 1960s (before the discovery of the cosmic background radiation) the prospect that the Universe emerged from a singularity bothered several Soviet physicists because it seemed distinctly unphysical.[2] Singularities are points where a mathematical function becomes infinite or undefined. That is fine in mathematics, where the function can be an abstract construction, but in physics, where the

3.2 An Early De Sitter Phase

function is an actual physical quantity, such as energy density or the curvature of space–time, it is problematic. One possible way around this problem not involving speculative new physics at the Planck scale is to suppose that the initial Universe is not dominated by actual matter or radiation, but is in a vacuum state with negative pressure; i.e., pressure is equal to the negative of the energy density ($p = -\rho c^2$). This strange fluid overcomes attractive gravity and drives an exponential expansion. Moreover, because of the unusual equation of state, the density does not dilute with the expansion of the Universe, but it remains constant. This, in fact, is a re-introduction of Einstein's cosmological constant and the resulting expanding matter-free universe is that of de Sitter; so the proposal would be that the world is initially in a de Sitter phase. Going back in time, this means that the density does not increase – there is no singularity.

The idea was proposed by a young physicist, Erast Gliner (1965, 1970), who thought that it would be an elegant mechanism of avoiding the singularity, but he suggested no specific mechanism for initiating or ending the exponential expansion. Gilner's proposal was briefly taken up by Andrei Sakharov in 1970. Sakharov tried several equations of state for the vacuum fluid (the relation between pressure and density) and even suggested that density fluctuations might be generated in the rapid expansion – the fluctuations that are currently observed in the large-scale distribution of galaxies (in retrospect, a remarkably precocious suggestion). But in these early models there was no clear physical motivation for the de Sitter phase. An initial step in this direction had been taken in 1967 when Yakov Zeldovich realized the connection between the cosmological constant and the zero-point energy of quantum fields (in quantum mechanics, if one cools a material to absolute zero there remains a residual energy density – the zero-point energy that in the context of Einstein's theory should also gravitate, or rather anti-gravitate). This suggests a possible mechanism for a large vacuum energy density connected with a quantum field at high energy.

This search for a physical motivation of the initial de Sitter phase was carried further by Alexei Starobinsky, who in 1979 applied new concepts in quantum gravity (such as Hawking radiation) to demonstrate how particles may be created in strong gravitational fields, suggesting a mechanism for a transition between the de Sitter phase and the subsequent matter–radiation dominated FLRW expansion of the Universe. Starobinsky also pointed out a possible observational test for the initial de Sitter phase: an exponential expansion generates not only density fluctuations but also gravitational radiation. There should be a sea of low-frequency gravitational waves from an epoch much earlier than that of the observed microwave background.

The proposal of an early phase of exponential expansion arose not from any experimental or observational contradiction with the Big Bang; the motivation for the early work was not phenomenological but aesthetic: the avoidance of the singularity. Considerations of naturalness – the quest to avoid fine-tuning of initial conditions – were of secondary importance, although the advantages of de Sitter expansion in this regard were pointed out by Starobinsky, who realized that such a de Sitter phase would drive the Universe to very nearly zero curvature. But in 1980, interest in the concept of an initial exponential expansion took off when several young physicists proposed a physical basis for the de Sitter phase – a basis inherent in modern attempts to push high-energy particle physics beyond its very successful standard model. Then the emphasis shifted more to the problems of fine-tuning rather than the avoidance of singularities. Most conspicuous among the newcomers were Alan Guth in the United States and Andrei Linde in the Soviet Union. It was Guth who, in a brilliant stroke of genius in nomenclature, invented the term "inflation."[3]

3.3 The Physical Basis of Inflation

There are four known fundamental forces of nature: the strong nuclear force that holds atomic nuclei together, the weak nuclear force that mediates certain interactions between sub-atomic particles (such as beta decay that transforms the neutron into a proton, an electron and an anti-neutrino), the more familiar electromagnetic force of attraction between opposite charges and repulsion between similar charges, and the gravitational force. The basic idea behind inflation is that at least the first three of these forces become unified into one force at very high energies – in excess of 10^{15} GeV (10^{24} electron volts). This energy is equivalent to tossing the hypothetical bowling ball at a speed of only 1000 km/hour, so a factor of 100 slower than our Planck-scale bowling ball. Moving back in time, individual particles or photons would achieve this energy when the Universe is 10^{-34} seconds old, so in this sense the early Universe provides the ultimate particle accelerator (the "poor man's accelerator" in the words of Zeldovich) for testing theories like Grand Unification. The point is that the Universe is much simpler at such very high energies (temperatures) or at very early times when there is only one force between particles. With respect to cosmology, the important aspect of Grand Unification is the existence of a cosmic field, designated ϕ and associated with the "grand-unified force". This is not the field that mediates or "carries" the grand-unified force, but rather the field that gives mass to the particle that mediates the force (in that sense it is like the "Higgs" field). Due to the interaction of ϕ with itself, it possesses an energy density – a potential energy that permeates the Universe and at high energies contributes as an energy density of the vacuum.

3.3 The Physical Basis of Inflation

In the Grand Unification scenario, it is this vacuum energy density that creates the accelerating expansion of the Universe: the ϕ field is the "inflaton" – the field with a background energy density, a potential energy, that drives the accelerated expansion. This potential energy can be described as a cosmological constant in Einstein's equations, and provides a microphysical basis for the early de Sitter phase as well as the necessary exit from the de Sitter phase.

This potential, the energy density as a function of the scalar-field strength, has a very symmetric form at energies higher than the Grand Unification scale; at ages less than 10^{-34} seconds, there is only one vacuum, or lowest-energy, state, with an energy density less than or comparable to the thermal energy of the particles comprising the Universe. So the potential plays no role in driving the expansion of the Universe at these high energies; the expansion is decelerated, as in Friedmann expansion, by the gravity of particles and photons. But at lower energy densities (lower than the Grand Unification scale), the potential is less symmetric with two vacuum states – one, a false vacuum, corresponds to the energy density of unification (at $\phi = 0$) and the other, the true vacuum, lies at a much lower energy density (a smaller cosmological constant) where $\phi \approx 10^{15}$ GeV (Figure 3.1). So the Universe during its expansion can get hung up in the false vacuum and exhibit an exponential expansion caused by the huge cosmological constant. Eventually, after expanding by 70 or 80 factors of two, supercooling and diluting to essentially zero matter density, the field decays into the true vacuum with the energy being released into new particles (the Universe is essentially recreated by this reheating) and the traditional Friedmann expansion of the Universe continues. The inflationary period lasts for perhaps 10^{-32} seconds, so it is a short, but important, event. By providing the enormous expansion and the flood of new particles that fill the Universe all from the vacuum, inflation is, in the words of Guth, "the ultimate free lunch."

Inflation – the enormous expansion of space followed by reheating – obliterates, in principle, the need for fine-tuning of initial conditions. It is equivalent to inflating a pebble one millimeter in diameter to a size of 200 billion light years; any irregularities on the surface of the pebble are wiped out by this expansion and its initially curved surface appears to be extremely flat; that is to say, inflation drives the curvature of the Universe to very nearly zero, or the density to very nearly the critical value ($\Omega = 1$). Moreover, the size of a causally connected region, the horizon, is inflated from 10^{-25} centimeters to a scale of 3 kilometers, which by a redshift of 1100 when the black body photons last interacted with matter, will have expanded to 10 billion light years. This means that the entire opaque shell emitting the CMB photons (380 000 light years in size) would have been in causal contact before inflation. It is no mystery then that the Universe appears to be so isotropic. A region much larger than the entire observable Universe is in causal contact before inflation.

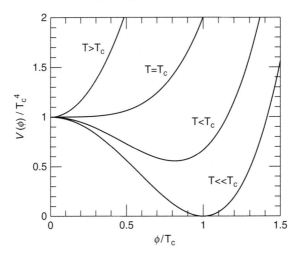

Figure 3.1. The effective potential in the original inflationary scenario for different values of the temperature of the background medium. The horizontal axis is the value of the scalar field in units of the critical temperature – the energy at which the vacuum begins to dominate the energy density of the Universe. The vertical axis is the energy density in the potential of the scalar field in units of the critical energy to the fourth power. The different curves show the effective potential at various values of the energy of the thermal background in units of T_c, a critical temperature corresponding to the energy of unification, supposedly 10^{15} GeV. At high energies ($T > T_c$) the potential is very symmetric with one vacuum state. At low energies the potential has a lower symmetry with two vacuum states: a false vacuum at $\phi = 0$ and a true vacuum at $\phi = T_c$. The Universe is hung for some time in the false vacuum, and this drives the exponential expansion – the inflation.

But inflation does more. During the inflation epoch, small quantum fluctuations are generated – fluctuations of the right form to create the observed structure in the Universe. These fluctuations would have been pushed by the exponential expansion beyond the horizon, so they would appear to be causally disconnected. Because there is (almost) no physical length scale connected with exponential expansion, the fluctuations would have been (almost) scale-free – the amplitude of the fluctuations would not depend upon their size. The fluctuations so produced would be *adiabatic*. That means that, after reheating, all particle species and photons would reflect the same fractional density variations. The temperature of the mixture would vary with density, and as we shall see, this is entirely consistent with the actual observed fluctuations in the CMB. So inflation creates density fluctuations of the right form and kind to later evolve into the observed structure by gravitational instability.

Moreover, these fluctuations come in two flavors: the basic tiny variations in the overall density are generated by quantum fluctuations of the scalar field that drives inflation – the inflaton. These are scalar fluctuations. And then there are quantum fluctuations in the underlying gravitational field that describe the shape of the four-dimensional space–time. These are, in effect, gravitational waves, or *tensor* fluctuations. It is actually possible to distinguish these two forms of fluctuations because they produce distinct patterns of polarization of the CMB. If inflation actually occurred, not only scalar but also tensor fluctuations must be present at some level. The detection of tensor fluctuations, this sea of gravitational waves, would, in a sense, provide the "smoking gun" evidence that inflation is a real event in the early history of the Universe.

Inflation accomplishes quite a bit, and it arises from basic physics in the early Universe rather than fine-tuning of initial conditions. It is an idea based upon the supposition that the world possesses a symmetry at high energies that is broken at the lower energies of the present Universe, so the concept is actually quite Platonic – at a deeper level the world is more symmetric than its realization in our low-energy experience.

In fact, inflation as an idea has evolved over the decades since its inception and has become considerably more baroque. The original theoretical basis for inflation – the Grand Unified Theory – has been dismissed because of the absence of the predicted decay of protons (with a lifetime predicted to be 10^{31} years). Other theoretical bases, such as supersymmetric inflation, have emerged with different forms of the potential energy designed to provide certain necessary attributes – for example, the necessary reheating of the Universe after the supercooling occurring during the exponential expansion (more on supersymmetry, a possible extension to the standard model of particle physics, will follow later). It can be justifiably argued that inflation is not in fact a theory (there is no proper microphysical theory behind the idea) but a paradigm – a wish list of what we would like inflation to accomplish. Andrei Linde, one of the founding fathers of inflation, has argued that inflation occurs continuously – that some patches of the Universe will not inflate, some will inflate forever, and some will inflate by just the right amount to produce us – so the idea of inflation has evolved to a multiverse concept. However, in spite of its varied manifestations and lack of a definite microphysical basis, the idea has remained potent because of its perceived predictive power – scale-free adiabatic fluctuations, a nearly flat universe (with density near the critical density) and a prediction of gravitational waves generated during inflation, waves that may eventually be observed in the pattern of polarization of the CMB.

As a concept, inflation has had a powerful impact; it is implicitly accepted by most cosmologists in spite of the absence of a definite microphysical basis. Most often, those properties which inflation was invented to explain – isotropy of the Universe, near-zero curvature, fluctuations that form the observed structure – are assumed to be absolutely established by the occurrence of inflation; there can be no doubt that the density of the Universe is equal to the critical density because inflation requires it. The paradigm has its critics (for example, Paul Steinhardt, one of the original pioneers of inflation, who now says that it should be called the "inflation myth"). It is not evidently falsifiable; even the failure to find the predicted gravitational waves would fail to falsify the idea since the magnitude of these fluctuations can always be pushed below the limit of detectability. The absence of falsifiability is carried further in its multiverse extreme in which anything, and therefore nothing, can be predicted. Yet there is no denying the power of the concept and its underlying motivation: to explain the apparent fine-tuning of the Universe through natural physical processes. And, as we shall see in the next chapter, the detectability of sound waves in the coupled baryon–photon plasma that existed before redshift of 1000 constitutes a non-trivial argument that something very much like inflation occurred in the early Universe.

As noted in the Chapter 1, there are two trends in mythological and scientific creation scenarios: the Universe is finite in time (or semi-finite) with a moment of creation in the past, and evolution – an ever changing aspect – since that moment; or the Universe is infinite and eternally constant in its average attributes. This dichotomy was reflected in the 1950s in the heated dialectic between the Steady State and Big Bang cosmologies and finally resolved in favor of the Big Bang by the discovery of the CMB. But we can recognize similarities between the inflationary universe and the Steady State universe. The Steady State required the presence of a constant Hubble parameter, an exponentially expanding universe with no preferred scale, a constant energy-density background and, therefore, particle creation. Inflation, at least during the inflationary period, also proposes a constant Hubble parameter, an exponential expansion and a constant energy density (albeit in the vacuum). The essential difference is that the inflationary epoch is a temporary, extremely short-lived phase at the origin of the Universe, with particle creation occurring from decay of the false vacuum at the end of inflation. But even this distinction blurs in the context of the multiverse extension of inflation – eternal inflation – in which the multiverse is again infinite in time and space and constant in the sense of its ongoing mixture of big and little bangs. So with regard to the history of ideas we once again discern a pattern – conflicting models, intense dialectic, resolution in favor of one model, but finally synthesis.

4

Precision Cosmology

4.1 Standard CDM Cosmology

Shortly after the discovery of the CMB, Jim Peebles, closely followed by Robert Wagoner, William Fowler and Fred Hoyle, realized that the measured abundances of those light elements produced in the first few minutes of the Big Bang were sensitive probes of the density of the Universe, or at least the density of ordinary protons and neutrons that comprise most of the matter that we can directly observe – baryonic matter.[1] This is particularly true of deuterium (a hydrogen nucleus with a neutron as well as a proton) and helium-3 (a helium nucleus consisting of two protons and only one neutron, rather than the usual two). If the primordial cosmic abundances of these isotopes could be precisely measured, then this would comprise a probe on the number of baryons relative to the known number of black-body photons – usually designated η; that is to say, these measured abundances would constitute a "baryometer."[2] The abundance of helium, the second most abundant element after hydrogen, is actually rather insensitive to the density of baryons, but is much more sensitive to the expansion rate of the Universe in the first few minutes after its origin. This depends upon the number of relativistic particle species such as neutrinos – the more species, the faster the expansion rate and the higher the helium abundance (recall that free neutrons have a lifetime of about 10 minutes; with a higher expansion rate more neutrons are available to form helium nuclei at temperatures low enough to avoid immediate photo-destruction). The expansion rate also depends upon more speculative possibilities such as deviations from standard gravity (i.e., a higher or lower constant of gravity in the early Universe). That is why it is said that the measured helium abundance is an effective chronometer.

Measuring the pristine abundances of helium or of the trace isotopes deuterium or helium-3 is a tricky business because of subsequent processing of these elements in stellar interiors – so-called "astration." Helium and helium-3, for example, are produced in stars but deuterium is destroyed. So how does one determine the primordial unprocessed abundances of these isotopes? For deuterium it is best to look at absorption lines in the light of distant (hence old) objects – the very luminous quasi-stellar objects associated with the accretion of massive black holes in the centers of galaxies formed relatively early in the history of the Universe. The absorption lines do not arise in the object itself but from intervening gas clouds along the line-of-sight to the quasi-stellar object. Judging from the very weak lines of elements heavier than helium, there has been little contamination of these clouds from gas cycled through stars – little astration – so the inferred abundance of deuterium is largely primordial. These deuterium lines are weak compared to the much stronger lines of normal hydrogen, so it is a difficult observation. Nonetheless, the results of observations and analysis have established that the implied density of baryons is significantly lower than the critical density for the flat Friedmann model – perhaps five percent of the critical density – so the relevant Friedmann model would apparently be the saddle-shaped open universe that expands forever.

This is problematic because a flat universe – a universe with a particle density very near the critical value – is the most natural in the context of the Friedmann equations (natural in the sense that if the density is critical, it is always critical), as well as being the property enforced by the inflationary paradigm. Yet the observed abundances of light isotopes imply that the density is only five percent of the critical density. How can this be reconciled with the flat universe at critical density, the model that is most consistent with naturalness and inflation?

There is a second difficulty with a low-density baryonic universe, which is an observational problem as well as a problem in principle; this concerns the growth of fluctuations and the formation of structure. Fluctuations in the baryonic matter can begin growing only after matter decouples from radiation at a redshift of 1100. Before this point the pressure of the coupled baryon–photon fluid, provided primarily by the radiation, is too high. Then, from decoupling to the present epoch, the fluctuations can grow by a factor that is roughly equal to redshift, roughly 1000, but this is only true in a universe with critical density. If the density is on the order of 0.05, as implied by the abundance of the light elements, then the fluctuations can only grow by a factor of 50. The existence of gravitationally bound structures such as galaxies and clusters of galaxies means that by the present epoch the density contrasts must be at least 100% or $\Delta\rho/\rho \approx 1$. In other words, initially at $z = 1100$ when growth could start, it would be necessary for $\Delta\rho/\rho \approx 0.02$. However, by the mid 1970s, this had already been ruled out by the absence of comparable temperature fluctuations in the CMB. It appeared to be impossible to form the observed structure in the Universe by the gravitational instability of baryonic matter alone.

The proposed resolution of this apparent conundrum lies in the fact that the nucleosynthetic limit on density applies to baryons – ordinary matter consisting of protons and neutrons. If there is other matter – non-baryonic matter – that interacts very weakly with baryons and photons and is non-relativistic at the epoch of nucleosynthesis (cold non-baryonic matter), then this limit does not apply. In the first place, it is possible to construct a flat $\Omega = 1$ universe with a low baryonic density if that universe is filled with non-baryonic dark matter. This leads to a faster growth rate for the fluctuations. In the second place, because such matter does not interact with photons, the growth of fluctuations in this dark component can begin much earlier than at $z = 1100$ when the photons decouple from the baryonic matter; the growth of fluctuations in the dark component begins when the matter first dominates over radiation at a redshift of about $10\,000$ if $\Omega_{\text{matter}} \approx 1$. So the growth of structure in the dark component gets a head start on that in the baryonic material. Later on, when the baryons do decouple from radiation, they fall into the pre-existing gravitational potential wells of the dark matter. Thus dark matter can provide the natural $\Omega = 1$ universe while promoting the formation of structure in the Universe.

By 1980, the concept of dark matter was already in place as a means of taming the instability of rotationally supported disk galaxies and providing the missing dynamical mass suggested by the observation of non-declining rotation curves of spiral galaxies, as well as by the observed dynamical discrepancy in clusters of galaxies (member galaxies are moving much too fast to be gravitationally bound if the mass is only in the form of the observed luminous matter).[3] Moreover, modern theories of particle physics were suggesting new particle candidates for the dark matter – particles beyond the known particles of the standard model of particle physics.

The standard model of particle physics has been very successful in explaining high-energy experiments and in predicting new particles that have subsequently been found. And yet, the model contains no particle candidates for the dark matter. To comprise the desired dark matter such candidates must be weakly interacting (for example, not possessing an electric charge as do the proton or electron), they must be stable (unlike the neutron that outside of an atomic nucleus decays in about ten minutes) and it should have a non-zero mass, ruling out, for example, gravitons that carry the gravitational force. The only possibility in the standard model menagerie is the neutrino, but standard neutrinos also do not work because their mass is too small to cluster on the scale of galaxies; they cannot possibly provide the dark matter content of galaxies.

But physicists have long suspected that the standard model is not the final word in particle physics theory; there is physics beyond the standard model. The leading contender for the more general theory is supersymmetry, which posits a symmetry between bosons and fermions. Sub-atomic particles have the property of spin (or angular momentum), and this spin is quantized; the angular momentum comes in integral or half-integral units of $h/2\pi$. The particles with integral spin are called bosons (like the photon with spin-1), and those with half-integral spin (like the electron) are fermions. In supersymmetry the proposal is that every fermion has a integral-spin partner and every boson has a half-integral spin partner; the theory effectively doubles the number of particles by introducing these "superparticles." But since none of these particles are actually observed, they must be massive and/or unstable with very short lifetimes – that is, all except the lowest-mass superpartner, which is a candidate for the dark matter particle.

Such a hypothetical massive particle would constitute dark matter that is cold in the sense that it would be non-relativistic when it decouples from photons and other relativistic species. This cold dark matter, CDM, would promote structure formation on all scales and make up the Newtonian mass budget of bound systems such as galaxies and clusters of galaxies. In the 1980s, these perceived properties led to the model of "standard cold dark matter" or SCDM: a model universe which is 95% dark matter and 5% baryonic matter.

The SCDM model universe has several attractive attributes. There are only two components of the Universe and the energy density of both declines in the same way as the Universe expands (as the inverse of the expanding volume). There are even theoretical motivations for the density parameter of dark matter being so nearly equal to one ($\Omega_{dm} \approx 1$). But rather quickly (in the early 1990s) the model was challenged by observations of the baryonic content of clusters of galaxies in comparison with the total dynamical mass.[4] In rich clusters, most of the baryonic content is in the form of hot gas. The measurement of gas temperature and the density of this hot gas can be achieved using X-ray observatories on platforms in orbit around the Earth. With these observations, and assuming hydrostatic equilibrium, one can calculate the dynamical mass of a cluster (including the dark component) as well as the total mass in hot gas. Numerical calculations indicate that clusters should sample a representative volume of the Universe; that is to say, the ratio of non-baryonic to baryonic mass should be close to the universal value – about twenty to one. But in fact, the discrepancy appears to be closer to six to one; in other words, the observations of clusters implied that there was insufficient non-baryonic dark matter to make $\Omega = 1$, i.e., a universe with zero curvature at critical density.

The perceived problem was resolved in 1998 with the discovery of the accelerated expansion of the Universe. Exploding stars of a certain type (type Ia supernovae) appear to be a rather precise standard candle (having a definite luminosity) at maximum brightness. Utilizing observations of distant supernovae (out to a redshift of one), two groups discovered that the supernovae appeared to be systematically under-luminous at higher redshift; this can be explained if the expansion of the Universe is accelerating, and, in the context of general relativity, this is possible if the Universe is currently dominated by a cosmological constant – Einstein's old fix for a static universe and de Sitter's original model for an empty expanding universe.[5] So there is another component of the Universe in addition to baryonic and dark matter – a vacuum energy density amounting to about 70% of the critical energy density of the Universe. The Universe now appears to consist of three components: ordinary baryons at 5%, dark matter at 25% and the energy of the vacuum, "dark energy," at 70%. The preferred model for the Universe is no longer that of "standard CDM," but is now ΛCDM, where Λ is the symbol for the cosmological constant.

4.2 Primordial Sound Waves

At the beginning of the new millennium, a major breakthrough in the study of the CMB was made possible by the placement of new sensitive microwave detectors with higher resolution on platforms at high altitude or in space: the observation of small-scale anisotropies in the microwave background. The COBE discovery of anisotropies in the CMB applied to large-scale fluctuations – much larger than those that formed the observed structure in the Universe. But starting with the *BOOMERANG* and *MAXIMA* balloon-borne experiments[6] in 2000, and proceeding onto the *WMAP*[7] and *Planck*[8] satellites three to thirteen years later, anisotropies down to arc minutes have been detected and mapped, and this has led to an astounding advance in defining the parameters of the standard ΛCDM model of the Universe. The discovery is all the more remarkable because the theory of the systematics of the fluctuations proceeded the actual observation. What has been seen on the sky are essentially sound waves in the photon–baryon fluid, or rather, the traces of these sound waves that are frozen into the CMB at the instant when these two components decouple from one another. These sound waves are, in the truest sense, the music of the spheres.[9]

The picture is basically this: In the very early Universe, inflation, that brief de Sitter phase of exponential expansion driven by vacuum energy, generates small fluctuations and drives them to sizes larger than the causal horizon. Significantly, inflation delivers growing fluctuations (both positive and negative), compressions and rarefactions; it establishes a temporal phase coherence – a coherence in time. Later on, when the usual Friedmann expansion of the Universe resumes, the horizon catches up with these fluctuations; this is because the horizon grows linearly with cosmic time (it is proportional to the age times the speed of light), but the universal scale factor which measures the increasing size of all lengths, including the size of fluctuations, grows directly with cosmic time to a smaller power (proportional to the square root of time when radiation dominates the Universe or to cosmic time to the two-thirds power when matter dominates). So at some point a fluctuation on a given scale re-enters the horizon and becomes, in effect, causally connected (happening sooner for smaller-scale fluctuations). When this occurs before decoupling of the baryons and photons (before recombination of the baryons), the Jeans length of the fluctuation, the minimum scale of a gravitationally unstable region, is formally larger than the horizon, so the fluctuation is too small to collapse under its own gravity; therefore it oscillates as a sound wave. But that is only true for fluctuations in the baryon–photon fluid; the cold dark matter has no restoring pressure, so positive fluctuations on all scales in this medium slowly begin to collapse. At an age of 380 000 years or a redshift of 1100, the baryons combine with electrons, become neutral and are free to collapse into the existing potential wells; the photons are released to stream directly to us 13.8 billion years later and reveal the pattern of acoustic oscillations.

These sound oscillations are basically similar to those with which we are familiar – longitudinal compressions and rarefactions propagating in the air around us. But primordial oscillations of certain definite wavelengths are distinctly visible in the microwave sky. Primary of these are those oscillations which have just had time to compress once by the epoch of decoupling – that is, oscillations with a half-wavelength equal to the distance that sound has travelled since the beginning of the Universe – the sound horizon. Because the sound velocity in a baryon–photon coupled fluid (with pressure dominated by photons) is the speed of light divided by the square root of three ($c/\sqrt{3}$), the sound horizon is proportional to the causal horizon. We could call this special oscillation the fundamental wavelength. Then there are the oscillations which have just had time to compress once and re-expand to maximum rarefaction within the age of the Universe – the first overtone. Next is that oscillation which has compressed, re-expanded, and compressed a second time – the second overtone, and so forth.

What we actually see in observations such as those made by the *Planck* satellite is a map of the entire sky showing brighter and darker spots as in Figure 4.1.[10] This pattern results because the intensity of the fluctuations in the background radiation is equal to the degree of compression, or rarefaction, in the acoustic oscillations. This map is impressive considering that the fluctuations in intensity are on the order of 5/100 000 of the overall brightness of the CMB, but it is not very quantitative. It is more informative to decompose this pattern of warmer and cooler spots into a series of multipole moments, an angular power spectrum, because that reveals the power in the temperature fluctuations – basically $(\Delta T)^2$ – on various angular scales. Because the screen we are observing (with a redshift of 1100) is at a fixed distance, the angular scale can be translated to linear scales, and any peaks that we observe in the power spectrum would correspond to oscillations with a

4.2 Primordial Sound Waves

Figure 4.1. An all-sky map of the intensity of the CMB as observed by the *Planck* satellite; this reveals the tiny temperature variations ($\Delta T \approx 10^{-5}$) in the background radiation. The small red and blue patches (warm and cold spots smaller than an angular size of one degree on the sky) reflect the acoustic oscillations. See rear cover illustration for a full-color version.

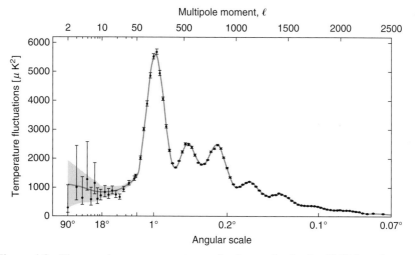

Figure 4.2. The angular power spectrum of anisotropies in the CMB based upon the *Planck* observations shown in Figure 4.1. The horizontal axis (lower label) is the angular scale of the fluctuations. The upper label corresponds to the index of the multipole moment. The vertical axis is the temperature variation in units of micro-degrees squared. The peaks show the acoustic oscillations at preferred wavelengths corresponding to the fundamental oscillation at one half of the sound horizon and its higher harmonics.

definite wavelength. Such a power spectrum is shown in Figure 4.2 for the recent observations made from the *Planck* satellite.

This figure shows the power in the temperature fluctuations plotted as a function of angular scale from large (left) to small (right); that is to say, the first peak on the left has the largest angular scale corresponding to somewhat less than one degree on the sky (the top axis is labeled in terms of the multipole moment, l). Of course, it also has the largest linear scale and that would correspond to a region that has, by now, expanded to roughly 500 million light years. In fact, the series of peaks reveals the fundamental mode (the first peak) and the various overtones of the sound waves in the baryon–photon plasma, and contains a wealth of information. First of all, in wavelength the peaks form a harmonic progression – almost. The form of a true harmonic progression is l_0/n, where l_0 is a fundamental length scale and n is the number of the peak. This would be expected in the context of sound waves emerging in the primordial baryon–photon plasma. Actually, the first peak does not fit into this sequence very well – the interval in angle in the first two peaks is a bit too large; more precisely the distribution of peaks, in wavelength measured in millions of light years, is of the form $1/[a+(n-1)b]$. The position of the first peak is, in effect, a standard yard stick – it reflects the distance that sound has traveled since the Big Bang – the sound horizon or about 500 million light years (expanded to the present Universe) – and therefore its angular size is sensitive to the topology of the Universe. It turns out that the Universe is very nearly flat with $\Omega = 1$, as is provided by the inflationary model.

It seems remarkable that the pattern of peaks is so nearly harmonic given the fact that these are not pure sound waves in a simple plasma; there are complicating effects. First of all, there are two components of the plasma – baryons, which provide inertia, and photons, which provide the restoring pressure. These oscillations are taking place in an evolving background in which the density and temperature are continually decreasing; moreover, there is the presence of pre-existing potential wells provided by the dark matter component that does not take part in the oscillations. These effects provide not only deviations from a pure harmonic sequence but also varying amplitudes for the harmonic peaks – different for compressions (odd-numbered peaks) and rarefactions (even-numbered peaks). For very high-order overtones (small wavelength), photons can leak out of the density extremes and this reduces the overall amplitude of the peaks. All of these effects reflect the imprint of baryons, photons and dark matter, so the precise pattern of the harmonic peaks, positions and amplitudes, are probes of the relative contribution of these components. Quite literally, the history and composition of the Universe is written in the book of the sky.

Like ancient hieroglyphs, the language of the book of the sky is complicated, but in this case the Rosetta Stone is provided by computer programs which include all of the effects mentioned above. Basically, one adjusts half a dozen free parameters of the model, such as the curvature of space–time and the abundances of the different constituents, and tries to fit the angular power spectrum shown in Figure 4.2. The fit is remarkably good (solid curve) and the parameters are consistent with other observations. First of all, the fraction of the total present energy density provided by baryons is about 5% – within the errors of the determination provided by the analysis of the primordial composition of light isotopes such as deuterium. These two determinations – primordial nucleosynthesis and acoustic oscillations – rely on different physics at very different epochs in the evolution of the Universe. Nucleosynthesis occurs when the Universe is a few minutes old and involves processes in atomic nuclei such as beta decay and neutron absorption cross sections. Baryons affect the acoustic oscillations primarily by adding inertia to the compression phase (reflected in relatively lower peak values in the rarefaction phase most noticeable in the second peak). The fact that these two results agree within the errors (resulting primarily from the deuterium abundance determination) is encouraging for the overall paradigm.

And then there is the dark matter, which forms the potentials for the oscillating plasma. This is detectable as a large height for the third peak relative to the second peak, and results from deeper compression in the potential well due to the baryon loading described above. Fitting to the details yields a contribution of about 25% of the dark matter component to the present total energy density of the Universe. But, of course, the position of the first peak implies that the Universe is flat and therefore at the critical density. The total matter contribution (baryonic plus non-baryonic) is 30%, so the remainder must be in dark energy – presumably a cosmological constant. This is completely consistent with the late-time acceleration of the Hubble expansion evidenced by the distant supernovae. The consistency of higher peaks (out to at least seven) with this dark matter abundance also implies that the dark component is reasonably cold – that it can form structures down to at least the scale of galaxy clusters.

The very existence of harmonic peaks constitutes non-trivial support for the inflationary paradigm, but this is a subtle point that requires a bit of thought. When a piano string is struck we hear harmonic oscillations because the string is connected at both ends to the rigid piano board; the ends of the string are constrained to be nodes or zero points of the oscillations. Thus, when the string is struck by the hammer, we can hear the fundamental tone with wavelength $\lambda = 2L$, where L is the length of the string, as well as overtones of $\lambda = 2L/n$, where n is an integer greater than one. But what is the equivalent process in the primordial baryon–photon plasma? What is the mechanism that ties down the strings and creates a preference for fundamental and harmonic overtones in the primordial plasma oscillations?

It is basically cosmic time that ties down the ends of the strings of this enormous piano in the sky. At one end, a fluctuation enters the horizon and begins oscillating; at the other end, oscillations cease when the baryons and photons decouple. We preferentially see those fluctuations that oscillate a half integral number of times

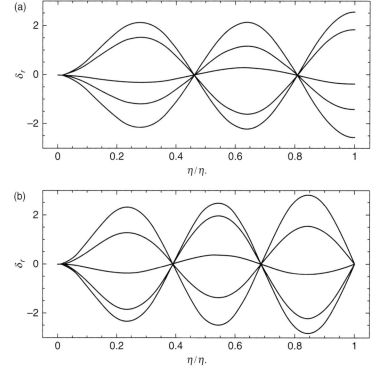

Figure 4.3. The phase focussing of fluctuations during the period of time when they oscillate as sound waves in the coupled baryon–photon plasma. The horizontal axis is basically cosmic time (a special sort of time that takes into account the expansion of the Universe) and the vertical axis is an arbitrary measure of the amplitude of the fluctuations. The upper and lower panels illustrate an ensemble of fluctuations of two different wavelengths. The fluctuations are all growing when they enter the horizon (near time zero) as delivered by inflation. The fluctuations of any given scale reach the node or zero point of their oscillations at the same cosmic time. This effectively causes the observed harmonic pattern of acoustic oscillations. The figure is from the paper of Albrecht et al.[12]

during this period between the epochs when they become causally connected and when they cease due to decoupling.

However, there is more to it. On the scale of the acoustic oscillations there are many fluctuations all over the sky of one specific size or wavelength. And yet, the amplitude of these oscillations, positive or negative, seems to be increasing when they enter the horizon; they are growing fluctuations. There is a degree of temporal phase coherence that would not be expected if the waves developed by some *in situ* causal mechanism. As is shown in Figure 4.3, there is a "phase focussing;" the fluctuations of a specific wavelength over the entire observable Universe reach the node or zero point of their oscillations at the same cosmic time (of course, they also reach the extremum of their cycle – maximum compression or rarefaction – at the same cosmic time). It is as though they are struck by some universal hammer at the same instant – even though they are separated by distances across the sky suggesting that there can be no causal contact.

That universal hammer is inflation – it is a mechanism that can coordinate the phases of these widely separated oscillations by creating growing fluctuations of a given scale everywhere at the same instant of time.[11] Of course, this could also be done by postulating a non-causal temporal phase coherence as an initial condition, but this would seem quite strange in a naturalistic world. So, independently of the microphysics of inflation (whatever field and potential energy drives inflation), the very existence of preferred peaks in the power spectrum of the anisotropies provides strong evidence that the Universe, very soon after the Big Bang, experienced a short de Sitter phase of exponential expansion driven by the vacuum energy of the Universe.[12]

4.3 The ΛCDM Paradigm

All in all, it appears that the small-scale anisotropies in the CMB reveal, in terms of the acoustic oscillations in the primordial baryon–photon plasma, a consistent picture of the Universe as well as a precise determination of its geometry and composition. It is the greatest success of modern physical cosmology. The pattern in these small anisotropies is, in effect, the trace of the Universe, the sign or evidence of its early evolution and composition.

Why then, should there be any doubt about the ΛCDM cosmology? As I suggested in the introduction, the doubt arises primarily because the natures of the principal components of the Universe – dark matter and dark energy – are mysterious. With respect to the dark matter, there are speculations upon the nature of these hypothetical particles involving extensions of the standard model of particle physics, but these remain speculations. Until the particles are identified in the laboratory, they can never be more than hypothetical constructs. The fluid of dark matter constitutes an aether for which there is no independent evidence. The nature of dark energy is likewise speculative. Is it a mere constant in the Einstein field equation? Is it truly constant or does it evolve with cosmic time? Should it be identified with the potential of an additional field? If so, we might expect to see evidence of this field in other phenomenology – a fifth force perhaps. Does the dark energy represent the zero-point energy of some quantum field? Why then does it have this unnaturally small value? These are fundamental puzzling questions in the context of ΛCDM cosmology.

The parameters of the Friedmann–de Sitter model may be precisely determined by the fitting to the angular power spectrum of the CMB anisotropies, but the unusual composition suggests that those models may not be valid. Perhaps general relativity is not applicable on cosmic scales. Perhaps there are additional cosmic fields that are not evident locally. Significantly, there are problems with the perceived properties of cold dark matter in connection with galaxy phenomenology. This phenomenology appears to point to a preferred constant of acceleration in the Universe which has no natural explanation in the context of ΛCDM. These problems suggest that, in spite of the success of the paradigm on cosmological scales, the picture is seriously incomplete. I will discuss these issues in later chapters, but first we will consider the overall consistency of the standard paradigm strictly in terms of cosmology.

5

The Concordance Model

5.1 Consistency

In 1995, before the observations of the anisotropies in the CMB offered such a precise view of the Universe at the epoch of decoupling, and before the discovery of the accelerated expansion of the Universe through observations of distant supernovae, Jerry Ostriker and Paul Steinhardt in a letter to the journal *Nature*[1] discussed the existing constraints on cosmological parameters. These were constraints arising from several distinct considerations: limits on the age of the Universe from globular-cluster lifetimes, measurements of the Hubble parameter H_0 using the Hubble Space Telescope study of classical Cepheid variables and supernovae, the contribution of baryons to the mass budget of the Universe based upon the measured primordial abundances of light isotopes in the context of Big Bang nucleosynthesis, the measured and assumed universal ratio of baryons to dark matter in clusters of galaxies, and the total matter density required to form the observed structure in the Universe. They assumed that the properties required by inflation were absolutely valid: a flat universe in which the total Ω, including matter, radiation and vacuum energy, is equal to one and in which the primordial fluctuations were scale-free. They concluded that the model in *concordance* with all of these constraints was a low-density world ($\Omega_m \approx 0.3$) dominated by a cosmological constant ($\Omega_\Lambda \approx 0.7$). In fact, it turned out that this *concordance model* was that which is supported, with much higher precision, by the subsequent observations of anisotropies in the CMB – altogether a remarkable success of inductive logic. The expansion and composition history of the concordance model are shown in Figure 5.1. It seems appropriate to refer to this model as the Friedmann–de Sitter model: it was pure Friedmann but is now becoming de Sitter.

Now we find ourselves in the situation where the primary empirical input to cosmology arises from the observations of the CMB. Although this is to be expected given the precision of the results, it is not altogether desirable to rely

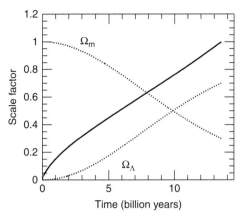

Figure 5.1. The expansion and composition of the concordance model. The solid curve shows the development of the scale factor with cosmic time (note that the cosmological redshift (z) is related to the expansion factor (a) as $z = 1/a - 1$). In the last three to four billion years, the accelerated expansion is evident. The contribution of cold matter (dark and baryonic) and that of the cosmological constant to the energy density of the Universe is shown in terms of Ω by the dotted curves. It is clear that the cosmological term begins to dominate over the matter at a redshift of about 0.4, coinciding with the observed accelerated expansion. It is also evident that we are currently in an epoch in which the fractional energy density in matter and in the cosmological term are nearly equal and are changing relatively rapidly.

upon one phenomenon arising at one cosmic epoch to set the entire cosmological scenario. It would be as though geologists and paleontologists were to rely upon the continental formations, the composition of the atmosphere, ocean and rocks, and the dominant fossils that characterize some early geologic period, the Cambrian for example, to describe the geological and biological history and subsequent evolution of the Earth. The Earth presented a very different aspect 500 million years ago: there were fewer continental land masses and these were quite barren; there was much more carbon dioxide and less oxygen in the atmosphere; plants were generally microbial and larger multicellular animals were mostly arthropods. It would be impossible to predict the subsequent development of the planet from even a very complete picture taken during the Cambrian. It may be argued, with some justification, that this is a misplaced analogy because the early Universe is a more simple physical system than is the biosphere of a rocky, watery planet. However, the point remains that it is questionable to base the entire scientific view of cosmology primarily upon one single snapshot of a brief epoch as elucidated by a single phenomenon – the observation of photons that last interacted with matter at the epoch of decoupling when the Universe was 380 000 years old.

Cosmologists recognize this issue and do look for other independent cosmological markers. But there is a great deal of credibility in the present community of cosmologists and astronomers, with most researchers searching for evidence that supports rather than challenges the current paradigm (I will discuss this sociology later). But perhaps skepticism is a healthier scientific attitude; perhaps it is more meaningful to look for contradictions rather than concordance. Dialectic can be a more direct route to scientific progress.

In searching for discord, here I first consider any indications of inconsistency in different observations of the CMB. After all, there are now a number of maps of the microwave emission produced by different instruments with different sky coverage and resolution. It would be negative for the paradigm if the cosmology derived from these various data sets differed significantly.

Problems with the cosmological paradigm are likely to show up as inconsistencies in the values of the cosmological parameters derived by other observational techniques. Here I will discuss the overall consistency of the paradigm established by different considerations: that of primordial nucleosynthesis as well as the CMB in determining the contribution of baryons to the present density budget of the Universe, and that of the cosmological parameters implied by observations of distant supernovae compared with that derived by recent observations of the CMB anisotropies. I will consider the prediction of the phenomenon of "baryon acoustic oscillations" and its subsequent detection in the large-scale distribution of galaxies, and finally I will discuss the apparent association of the largest-scale anisotropies with a preferred direction in the principal orbital plane of the Solar System – the "axis of evil" – and its implications for the paradigm.

5.2 *WMAP* and *Planck*

The two most complete maps of the CMB anisotropies covering the whole sky are provided by *WMAP*, a satellite launched by NASA in 2001 with the initial results reported in 2003, and the *Planck* satellite, launched by the European Space Agency (ESA) in 2009 with first results in 2013. For *WMAP*, the microwave receiver had five bands with a maximum resolution of about 13 arc minutes corresponding to the position of the third peak in the power spectrum of acoustic oscillations. *Planck* had a maximum resolution of five arc minutes corresponding to the sixth peak of the acoustic oscillations. The sensitivity of the *Planck* receivers was typically several hundred times greater than that of *WMAP*. Because the two satellites had different systems with different detectors, and the observations were reduced differently, the consistency of the results constitute a crucial test of the cosmological paradigm.[2]

In the basic ΛCDM cosmological model, without any extra physics such as new neutrino types or evolving dark energy, there are six basic parameters – parameters that can be adjusted to achieve an optimal fit to the angular power spectrum. These are: (1) the density of cold dark matter in units of the critical density multiplied by the square of the present Hubble parameter, $\Omega_c h^2$, where h is the Hubble parameter in units of 100 km/s per Mpc; (2) the density of baryons in the same form, $\Omega_b h^2$; (3) the actual value of the Hubble parameter, H_0; (4) the form of the primordial density fluctuations characterized by a parameter n_s, where $n_s = 1$ corresponds to the scale-free distribution predicted (almost) by inflation; (5) the overall amplitude of the fluctuations, A; and (6) a parameter τ, which gives the probability of re-scattering of CMB photons after the neutral hydrogen becomes re-ionized following the formation of the first stars at a redshift near 10.

There is a 2.5% offset between the mean amplitude of *Planck* and that of *WMAP*, with the *Planck* power spectrum at high-order moments (smaller angular scales) being systematically higher. This is probably a question of calibration – setting the absolute level of the flux is not a simple matter. But comparing these parameters derived from the *Planck* data with those from the earlier *WMAP* data we find near agreement (within 1.5 standard deviations). The two instruments reveal basically the same cosmology when parameterized by the preferred ΛCDM model. The largest differences are in the contribution of cold dark matter to the composition of the Universe, where 27% of the Universe is dark matter with *Planck* and 23% is with *WMAP*. But much of this difference is due to the different values of the Hubble constant: 67 km/s per Mpc with *Planck* and 70 km/s per Mpc with *WMAP*. These differences are not significant, although the *Planck* value for the Hubble parameter does differ significantly from that determined by more local methods. In any case, the differences that exist between the two CMB satellites are insufficient to threaten the paradigm.

When additional parameters are added the fits do improve (as is expected). Examples of such additional parameters would be: the curvature parameter, $\Omega_k = 1 - \Omega_{\text{total}}$, which describes the current deviation from a flat universe (characterized by $\Omega_k = 0$); the number of relativistic particle species at the epoch of decoupling, N_r, which is generally attributed to extra neutrino types with mass less than 1 eV; and the equation-of-state parameter for the dark energy, w (the relation between pressure and density is $p = w\rho c^2$, where $w = -1$ corresponds to the cosmological constant). When such parameters are allowed to be free, it is found that they do assume unconventional values (with the exception of the curvature parameter, which is generally near zero within the statistical errors). For example, $N_r > 3$, suggesting the presence of extra neutrino types or "dark radiation," and $w < -1$, implying a dark-energy density that actually increases with the expansion of the Universe or "phantom" dark energy. But in no case does the improvement of the fit to the observed power spectrum provide significant evidence supporting deviations from the base cosmology with six parameters.

When comparing *Planck* and *WMAP*, in spite of small deviations in the fitted parameters, the acoustic peaks are at the same angular separations and have comparable relative amplitudes; this agreement adds considerable confidence that the two instruments are observing the same Universe as characterized by the concordance cosmology.

5.3 The Density of Baryons

The nucleosynthesis of the light elements – deuterium, helium-3, helium-4 and lithium – occurs when the Universe has existed for about three minutes and before most of the free neutrons are gone. At this epoch, the expansion rate and temperature history are independent of curvature or dark matter or dark energy; thus the predicted abundances of these light elements depend upon the number of baryons at this epoch – i.e., the ratio of baryons to photons – as well as any deviations from the standard expansion rate due, for example, to extra relativistic particle species such as a number of neutrino types in excess of the three types included in the standard model. The point is that in standard Big Bang nucleosynthesis (SBBN), without the assumption of new physics, there is one and only one parameter that sets the primordial abundances: $\eta = n_B/n_\gamma$, the number density of baryons to photons; given the known number density of photons from the Planck law as well as the Hubble parameter, η is directly related to the present density of baryons Ω_B.[3] So the measurement of primordial abundances of these four light elements allows one to determine the baryon density budget of the Universe.

The problem is that these elements are further processed in stellar interiors – their abundances can change due to production or destruction in stars or in the interstellar medium of galaxies due to interactions with cosmic rays (see Chapter 4). In particular, helium-4 is produced in the primary production of energy in stars, but deuterium and helium-3 are generally destroyed, as is lithium.[4] So it is important to observe regions that are as free as possible of astration, this subsequent evolution of the primordial abundances – regions such as the atmospheres of old stars, interstellar clouds with low abundances of heavy elements (indicating little astration), and spectral lines in distant and therefore ancient quasars. As is shown in Figure 5.2, the results of the measured abundances of helium-3 and -4 and of deuterium are generally consistent with primordial nucleosynthesis[5] and point to $\eta = 5.8 \times 10^{-10}$, from which we infer that $\Omega_B = 0.0432$. That is to say, baryons provide between 4% and 5% of the present density of the Universe.

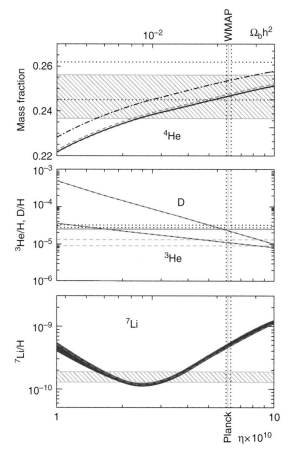

Figure 5.2. The curves show the predicted primordial abundances of helium (by mass fraction), deuterium, helium-3 and lithium (these last three are the abundances in terms of hydrogen) as a function of the baryon-to-photon ratio (η, lower axis, or $\Omega_b h^2$, upper axis). The vertical bands show the *WMAP* (dotted region) and *Planck* (solid band) results for η. The horizontal regions are the measured (and inferred) primordial abundances of these light isotopes. The agreement between the predictions of primordial nucleosynthesis, the CMB results and the measured deuterium is especially good, but the measured lithium is a factor of three too high. See Coc et al. [5].

The one problematic aspect of this analysis concerns lithium. The measured abundance of lithium in old stars is a factor of three or four below its predicted abundance with this preferred value of η, and this is well outside the statistical error bars. The problem could very well be with the stars. These stars are 10 billion years old, which would seem to provide ample opportunity for the depletion of lithium, although the precise mechanism is unknown.[6]

The acoustic oscillations in the baryon–photon fluid before hydrogen recombination also provide an estimate of the baryon density, but 380 000 years later and involving very different physical processes. Basically, the inertia in the baryons affects the measurable amplitude of the acoustic peaks, especially the even-numbered peaks that correspond to the rarefaction stage of the oscillations. This of course is completely independent of the neutron capture cross sections and reaction rates that determine the primordial abundances within a few minutes of the Big Bang, but the agreement of the two methods with respect to the baryon density is almost too good to be true (Figure 5.2). The acoustic oscillations imply that $\eta = 6.2 \times 10^{-10}$, corresponding to a baryon contribution of $\Omega_B = 0.046$, within 1.5 standard deviations of the SBBN result. If the number of neutrinos is taken to be about one more than the standard three (producing a higher expansion rate in the early Universe) then the agreement is even better, with nucleosynthesis yielding an η of 6.1 compared to the CMB value of 6.2 in units of 10^{-10}. This provides a non-trivial justification of both the standard Big Bang nucleosynthesis and the physical model behind the acoustic oscillations.

5.4 Supernova Cosmology

We saw in the previous chapter that type Ia supernovae (classified on the basis of their spectra) are characterized by a standard maximum luminosity, and when compared with their apparent brightness this allows an estimate of the distance.[7] The distance so determined, through use of a standard candle, is a special sort of distance – the luminosity distance – and the relation of this distance to redshift depends upon the cosmological model. If this dependence can be measured we can determine the model universe that we live in – specifically, the contribution of matter and vacuum energy density to the composition of the Universe.

The early results of observing distant type Ia supernovae revealed that these were too faint for a typical FLRW model, and this led to the conclusion, in 1998, that we live in a universe with accelerated expansion, which is possible in a universe dominated by a cosmological constant or a "fluid" with an unusual equation of state – that is to say, the energy density does not decrease with the expansion of the Universe as rapidly as for ordinary matter or radiation. The CMB anisotropies suggest that this dark energy component contributes about 70% of the present energy density of the Universe, so it is the dominant component of the world.

Now, more than 15 years after the discovery of the accelerated expansion and several continuing large supernova surveys[8], the data has achieved sufficient quantity and quality that useful constraints can be placed upon the cosmological

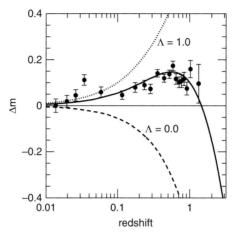

Figure 5.3. The Hubble diagram for type Ia supernovae normalized to an empty non-accelerating universe. The horizontal axis is the redshift and the vertical axis shows the difference in the observed magnitude and the expected magnitude for this model. The points are data binned in redshift from the supernova legacy survey (binned by Ned Wright). The dotted curve is the expectation for a flat universe dominated by a cosmological constant and the dashed line is the same for a matter-dominated universe. The solid curve is the concordance model with 70% dark energy and 30% dark matter.

models from supernovae alone. The results of one such survey are shown in Figure 5.3. This plot shows the difference in magnitude (apparent flux at maximum) between the observed supernovae binned in redshift intervals and a fiducial Friedmann model with zero matter and dark energy content. That is to say, a straight line would correspond to the $\Omega_{\text{total}} = 0$ Friedmann model. A positive apparent magnitude difference means that the observed supernovae are fainter than this model. The solid curve corresponds to the concordance model with 70% dark energy (in the form of a cosmological constant) and 30% matter. It is clear that the this curve reproduces the behavior of the observations; for redshifts between 0 and 0.5, the supernovae become dimmer than expected, with increasing redshift corresponding to the accelerated expansion and the dominance of the cosmological constant, and then, at redshifts approaching one, they become brighter, corresponding to matter domination and decelerated expansion. The other two curves correspond to topologically flat models dominated by a cosmological constant (dotted line) and entirely by matter (dashed line).

It is evident that the concordance model provides the adequate fit to the data. It is also clear that this mix of dark energy and dark matter produces the observed dependence of supernova peak magnitude on redshift. A model dominated by a cosmological constant would predict increasingly fainter supernovae with redshift. In the context of the Friedmann–de Sitter models, there must be a cold matter component (CDM and baryons) at a level (about 30%) that is in complete agreement with observations of the entirely independent CMB anisotropies.

The supernova data by themselves tend to favor a universe with an equation-of-state parameter less than -1, i.e., a universe dominated by "phantom" dark energy – a dark energy with a density that actually increases with the expansion of the Universe and that will end with a "big rip," an accelerated expansion that will eventually tear apart galaxies, stars and even atoms.[9] But when one adds the condition that the Universe should be flat, this preference vanishes.

Calibrating the supernova maximum luminosity by Cepheid variables in nearby galaxies, we may estimate the Hubble parameter. In this case we find a pronounced difference in the value given by the CMB anisotropies – the first serious tension between data sets.

5.5 Hubble Trouble?

One of the six parameters characterizing the standard model is the present value of the Hubble parameter, H_0. This sets the overall scale of the Universe – locally, as measured by the observed motion of relatively nearby galaxies, and on a cosmological scale by angular location of the first acoustic peak, approximately one degree (this has degeneracies with other parameters such as the curvature parameter and the dark matter density).

We have noted a slight but statistically insignificant disagreement between the *Planck* and *WMAP* results with respect to the Hubble parameter, where $H_0 = 67 \pm 1.2$ km/s per Mpc as determined by *Planck* and $H_0 = 70 \pm 2.2$ km/s per Mpc according to *WMAP*. This tension becomes greater when the *Planck* result is compared to more local determinations of the Hubble parameter – specifically to the Hubble Space Telescope calibration of the distance scale using type Ia supernovae calibrated by Cepheid variable stars in nearby galaxies: $H_0 = 73.8 \pm 2.4$ km/s per Mpc.[10]

This difference is statistically significant; the error bars of the two determinations do not overlap. But is the difference cosmologically significant? What could cause this tension between these two determinations of the Hubble parameter? It could be that there is a real difference in the Hubble parameter as measured locally via the supernovae and that measured cosmologically via the CMB anisotropies. If we, as observers, are near the center of a large empty region in the Universe – a void in the matter distribution – then the Universe would appear, locally, to be expanding faster than the true average value. If we are near the center of one of the largest empty regions possible in the context of ΛCDM then such a difference, about 10%, would be possible. So it is conceivable that the observed effect is a matter of cosmic variance.[11] However, it is also possible that the difference is real and due to a deviation from the base cosmology. An extra relativistic particle, a non-standard neutrino for example, with a mass less than a few tenths of an electron volt, would cause a greater expansion rate before the epoch of decoupling and therefore a smaller linear size for the sound horizon (less time for the sound wave to expand). Correcting for this effect would lead to a larger value of H_0 as estimated by the CMB anisotropies (primarily the angular scale of the first peak) and could bring the cosmological determination in line with the local measurement. Similarly, phantom dark energy promotes more consistency between the two determinations of H_0.

I have to admit that the evidence for these variations from the base cosmology is not compelling. Nonetheless, when two "precise" determinations of the Hubble parameter yield results that differ by more than the stated statistical errors, it is perhaps premature to claim that the results are in the category of "precision cosmology." There remains a systematic effect that is not yet fully appreciated.

5.6 Baryon Acoustic Oscillations

In everyday experience we often estimate, even unconsciously, the linear distance to an object by its apparent angular size. For example, the distance of an approaching human being can be judged by how large he or she appears to be, although there is always the possibility that the person considered is extraordinarily tall, leading to an underestimation of the true distance, or unusually short, and thus judged to be further away than in reality. However, on average, humans are about the same size, so if we observe a group of individuals we can make a reasonable guess as to their distance if we assume that they are all somewhat under two meters tall (unless, of course, the group consists of Dutch people). Humans would be an example of a "statistical standard ruler" (although we always have to be careful about systematic errors, such as that resulting from the Dutch selection effect).

The use of standard rulers has been an important distance indicator in astronomy, particularly for traditional observational cosmology. For example, assuming that compact radio sources associated with active galactic nuclei constitute a statistical standard ruler has led to conclusions on the matter content of the Universe.[12] However, now there is a new statistical standard ruler provided by cosmology in the sense that the primordial sound waves in the pre-decoupling Universe pick out a preferred length scale: the sound horizon. This length scale is given by the angular position of the first acoustic peak in the CMB anisotropy map and is predicted to be apparent in the large-scale distribution of galaxies – a more local indicator of the "baryon acoustic oscillations."[13] The value of this acoustic scale has expanded by now to roughly 500 million light years (the co-moving scale).

5.6 Baryon Acoustic Oscillations

To see how this length scale is imprinted upon the distribution of galaxies, let us consider a single spherically expanding sound wave that begins, after inflation, as an over-density – a compression. This means that, after the fluctuation re-enters the horizon, the pressure force acting upon the baryon–photon fluid is greater than the gravity force, so a shell of baryons and photons expands outward (this, of course, is an additional peculiar expansion over the usual Hubble flow). But the gas pressure does not act upon the dark matter, so the dark matter concentration remains in the center of the spherical region – it does not expand outward with the baryon–photon fluid. Then, at the epoch of decoupling, the shell of baryons has reached its maximum extent, the sound horizon, and the photons are released. There is no more pressure, and objects – galaxies – are free to form in the shell as well as in the center near the dark matter concentration. That means that, overall, there should be a concentration of galaxies at separations corresponding to the acoustic scale of 500 million light years (see Figure 5.4 for a cartoon of this process).

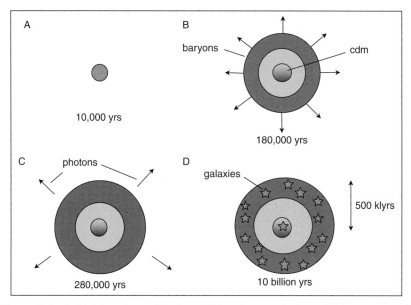

Figure 5.4. A highly schematic view of the time evolution of baryon acoustic oscillations. Frame A in the upper left shows an over-dense fluctuation after it enters the horizon at a cosmic time of 10 000 years. The baryons and the dark matter have not separated. After 180 000 years (frame B), the baryons are expanding due to the force of gas pressure (mostly provided by photons); the dark matter remains at the center because it is immune to this force. In frame C, decoupling has occurred and the photons are streaming away, leaving a shell of baryons. After 10 billion years (frame D), galaxies have formed in the shell (which has gravitationally attracted dark matter) as well as the center. There is a characteristic separation of 500 light years (co-moving) which is detectable in the large-scale distribution of galaxies – the baryon acoustic oscillation.

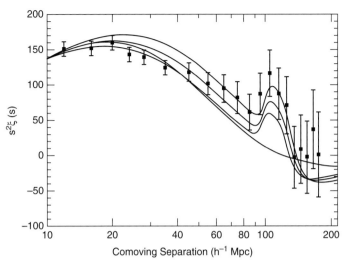

Figure 5.5. The galaxy correlation function as plotted against the co-moving separation from the Sloan Digital Sky Survey. The correlation function is the increased probability of finding two galaxies at the indicated separation and is an indication of the average run of galaxy density about a given galaxy. Here the correlation function is multiplied by the separation squared to better show the variations about its average behavior with distance. The separation is measured in megaparsecs (Mpc) divided by the Hubble parameter in units of 100 km/s per Mpc ($h = H_0/100 \approx 0.7$). The peak associated with the baryonic acoustic oscillation is evident at $100/h$ Mpc or at about 500 million light years; this is the scale corresponding to the first acoustic peak in the CMB angular power spectrum. The four curves, upper to lower, are associated with models having a matter density (baryons plus dark matter in terms of Ω_m) of 0.24, 0.26 and 0.28; the lowest curve is for the case of no baryon acoustic oscillations. Reproduced here with the permission of Daniel Eisenstein.

With the advent of large-scale galaxy surveys,[14] it has become possible to measure the average run of galaxy density about any given galaxy as a function of separation. The statistical measure is given by the galaxy correlation function, a measure of the probability of finding a galaxy at some distance, r, from another galaxy. We might expect the enhancement of galaxies at the baryon acoustic scale to be evident in the galaxy correlation as a function of distance, and indeed it is. This is shown in Figure 5.5, where the galaxy correlation is determined from the Sloan Digital Sky Survey (SDSS) of galaxy positions and redshifts over a large part of the sky. Here we see that there is an enhanced probability of finding galaxies at separations corresponding to the scale of the first acoustic peak.[15]

This observation appears to confirm the overall picture of the acoustic oscillations as revealed by the CMB anisotropies. Moreover, it is an independent confirmation related to the distribution of galaxies at much lower redshift than that of the CMB screen. When the position and amplitude of the enhancement in galaxy density is measured with greater sensitivity in upcoming galaxy surveys, this will allow investigation of the nature of the dark energy component of the Universe: does the density of dark energy remain constant with the expansion of the Universe (as a cosmological constant), or does it vary with redshift as one might expect from a dynamic field? So this is not only a confirmation of the overall picture, it is also a tool for further investigation of the nature of this mysterious constituent of the world.

5.7 The "Axis of Evil"

The idea that the Earth is not a special place in the Universe had its origin with the Copernican Revolution; Copernicus moved the Earth from the center of the Solar System (and presumably the Universe as a whole) to the position of a fairly small planet placed third out from the Sun – the true center. The history of astronomy since the sixteenth century has been a continuation of this process of relegating the Earth to a more obscure position in the world. The work of Harlow Shapley in 1910 moved the Sun from a position near the center of the Milky Way to a more remote position in the outer part of this great star system. Then Edwin Hubble demonstrated conclusively that the Milky Way Galaxy was only one such star system in a Universe of similar islands, some of which are much larger and more significant than the Milky Way. The message is – there is nothing special about us or our place in the Universe.

Given the direction of this movement – toward increasing obscurity in a large and mostly empty world – it therefore comes as somewhat of a surprise that the Universe, as reflected by the large-scale CMB anisotropies, appears to know about the Solar System, or vice versa. There are peculiar alignments between the largest-scale structure in the CMB intensity and the plane of the Earth's orbit about the Sun – the ecliptic. All of the planets in our system move more or less in this same plane; the same is true of smaller particles – the interplanetary dust that, by scattering sunlight, is responsible for the zodiacal light, a faint glow lying in a great circle on the sky corresponding to the ecliptic plane. At the distance of the great cloud of comets circling the Sun – the Oort cloud extending out to two light years – the ecliptic becomes inconspicuous; the outer comets are distributed in a more or less spherical region. Beyond these bound structures, at interstellar distances in the Galaxy, we expect no imprint of the orientation of our particular planetary system to remain, for example, in the distribution of nearby stars and certainly not in the distribution of nearby galaxies. Why then does the CMB arising at cosmological distances comparable to the present horizon (10 thousand million light years) exhibit alignments with this obscure planetary system?

When considering the large-scale alignments it is again useful to decompose the CMB intensity map into multipole moments – i.e., a mathematical series representing distribution of the microwave intensity on the sky. Each moment consists of a certain number of "spots," warm and cool regions arranged in patterns that become more complex for higher and higher multipoles. It is the low-order moments that reveal the largest-scale structure of the CMB brightness distribution. The first moment is the dipole with two lobes, one warm and one cool; this reflects the motion of the Sun, about 300 km/s, with respect to this cosmic frame.[16] And it is here that we encounter the first alignment: the axis of the lobes corresponding to the direction of the Sun's motion lies within 10 degrees of the ecliptic plane; this is to say, the Sun is moving with respect to the cosmic background almost within the orbital plane of its planetary system. This could easily be a matter of coincidence, but then comes a greater surprise. The next two multipole moments, the quadrupole and octupole moments, have four and six spots respectively. These spots, presumably arising from independent components of the CMB intensity distribution, lie almost in the same plane, and that plane is perpendicular to the ecliptic plane. This coincidence is due to large cool and warm and spots observed on opposite sides of the ecliptic plane: the orbital plane of the Earth falls between these most extreme regions of the CMB (see Figure 5.6). Moreover, the vector normal to this plane – that is, a line drawn perpendicular to the average plane containing the large spots of the quadrupole and octupole moments – is quite near the direction of the Sun's motion with respect to the CMB. This apparently preferred direction has been termed "the axis of evil."[17]

This effect was first noticed in the results from the WMAP satellite,[18] but more recently it has been confirmed by the *Planck* satellite. So the alignment is not

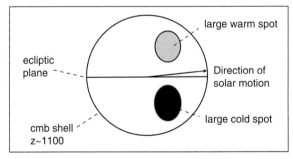

Figure 5.6. A schematic representation of the large-scale alignments in the CMB. The circle represents the shell at a redshift of 1100 emitting the CMB radiation. The superposition of the quadrupole and octupole moments of the observed anisotropy reproduces the large cool and warm spots on either side of the ecliptic plane and defines a preferred direction – the "axis of evil." The motion of the Sun with respect to the CMB lies within ten degrees of the ecliptic plane and near this preferred direction. See Copi et al.[18]

due to the particular detector, calibration or foreground-subtraction algorithm. It is apparently real and has a low probability of being accidental in a completely random universe – less than 0.1 percent.

What is the explanation for this apparent anomaly? It is certainly possible that the effect is completely coincidental; that we live in one part of a random universe where this particular large-scale alignment just happens to be evident. Computer simulations indicate that this is unlikely, but after all, it could be an aspect of the one-universe problem. The second possibility is that of an unsubtracted or unknown foreground contamination connected with the Solar System. Perhaps a currently unobserved large-scale distribution of dust about the Sun is emitting microwaves that contaminate the real signal of the background. The problem is that no one has constructed a convincing astrophysical model of how this might happen. The third possibility is that the effect is real and cosmological – the large hot and cold spots actually do lie in a preferred plane – but then its perpendicular orientation to the ecliptic plane in the Solar System is coincidental. The aspect of a preferred direction would violate the Cosmological Principle, but there is nothing sacred about this principle. We should also recall that the effect is small. The CMB is overwhelmingly isotropic. This peculiar anisotropy is occurring at the level of 0.00001 of the overall smooth, isotropic emission.

In any case, the primary cosmological information in the CMB arises from the small-scale structure – the structure in the brightness distribution on scales of one degree or smaller – that is present in the multipole moments greater than 100 (angular sizes of less than one degree). These reveal the presence of the acoustic oscillations which carry so much information about the composition and geometry of the Universe. Even if the large-scale alignments reflect some fundamental cosmological anomaly on the scale of the horizon (or for that matter, contamination on the scale of the Solar System), it appears unlikely that it would affect the regularities seen in the acoustic oscillations which are so well described in the context of the standard model. We would have to conclude that the concordance model is not seriously threatened by these large-scale alignments.

5.8 Summing Up: Absence of Discord

The standard cosmological paradigm, ΛCDM, provides a remarkably consistent picture of the world as probed by several independent tests. The primary of these is the observations of the CMB anisotropies and the acoustic peaks in the power spectrum. This set of data provides compelling evidence that the curvature of the Universe is near zero, that the initial fluctuations are adiabatic (all particle and photon species share in the variation of density) and that the fluctuations are nearly scale invariant (the amplitude does not depend upon the size). These properties are those expected to be delivered by the early de Sitter expansion of the Universe – the inflationary epoch.

Moreover, the detailed positions and amplitudes of the acoustic peaks tell us the composition of the Universe: the *Planck* data yields quite precisely 67.3% dark energy and 31.7% matter, of which only 4.9% is of the familiar baryonic form. This picture is confirmed by observations which are entirely independent of the CMB anisotropies. Considerations of primordial nucleosynthesis and the observed abundances of the light isotopes yields a density of baryonic matter consistent with that implied by the acoustic oscillations. The measurement of the apparent brightness of distant supernovae (the Hubble diagram), in providing direct evidence for the transition from decelerated to accelerated expansion at a redshift somewhat less than one, favors the same relative compositions of dark matter and dark energy implied by the acoustic oscillations. The baryonic acoustic oscillations observed in the large-scale distribution of galaxies establishes the existence of the acoustic scale which coincides with that observed in the CMB and again supports the overall picture of baryon–photon oscillations taking place in pre-existing potential wells that do not participate in the oscillations – wells that can only be formed by non-interacting dark matter.

That is not say that there are no tensions in the cosmological data. The significant difference between the Hubble parameter as measured from the CMB anisotropies and that implied by the supernova observations is a clear indication of systematic effects in one or both data sets. For example, perhaps the relatively local measurement from the supernovae is indicative of our position in a local void. The large-scale alignments in the low-order multipoles of the CMB anisotropies – alignments coinciding with the fundamental plane of the Solar System – call into question basic assumptions such as the Cosmological Principle (isotropy of the Universe) or even the Copernican Principle (no special position in the Universe). It may also be that, on a large scale, there are remaining problems with the subtraction of foreground emission, perhaps associated with the Solar System. But it is difficult to imagine that such problems would affect the consistent picture provided by the acoustic oscillations on a smaller scale.

Even a dedicated skeptic (as am I) would have to admit that that these independent lines of evidence – CMB anisotropies, the supernova Hubble diagram, the large-scale distribution of galaxies – provide strong evidence of the presence of dark energy and dark matter that produces predictable effects on a cosmological scale. The remaining fundamental problem is conceptual: What are these two substances? In particular, dark matter, as it is conceived to be, should cluster locally on the scale of the Milky Way, and therefore, at some level, it should be detectable by independent, non-astronomical means. So far, in spite of strenuous efforts and fond hopes, it has not made an appearance in terrestrial search experiments. Dark energy is more elusive: its density may be constant with universal expansion (the evidence suggests this) or it may be evolving and associated with an as yet undiscovered cosmic field. In either case it is strange and unexpected – an indication of possibly new gravitational physics.

Perhaps the standard paradigm is wrong. Perhaps there is no dark energy or matter, but the present Friedmann–de Sitter models are incomplete or incorrect. That is to say, perhaps the solution lies in a modification of general relativity on a cosmic scale. Then there must be a new and to some degree unconventional cosmology. But the bar for this new cosmology is very high: it must reproduce very detailed and consistent observations that are well accommodated in the context of the concordance cosmology.

6
Dark Energy

6.1 The Evidence for Dark Energy

If the creation and evolution of the Universe could be compared to baking a cake, then, in the context of ΛCDM plus *inflation*, the most important ingredient would be yeast – the substance causing an exponential expansion – a de Sitter phase. There are two such episodes of the dominance of yeast: the first is at the beginning, less than 10^{-32} seconds after the Big Bang. The modern creation story tells us that a fast-acting yeast causes the rapid expansion that inflates the Universe by a factor of possibly 10^{29}, wiping out any significant initial curvature (pushing the density toward the critical density), extending the causally connected region to well beyond the currently observable Universe, and creating the small fluctuations that become the observed structure. In the reheating that occurs after this inflation, the Universe is literally recreated; the initial conditions that applied before this episode, whatever they were, are irrelevant. Everything that the Universe is now observed to be, all of the existent ingredients in their present abundances, appeared out of the vacuum at the end of this initial yeast-dominated period. The microphysics of this event – the nature of the yeast – is unknown, but in a naturalistic world (one in which physical effects follow physical causes) there is substantial reason to believe that such an early de Sitter phase has actually occurred. We should keep in mind, however, that in the absence of primary evidence (such as gravitational-wave-like fluctuations) and without a specific physical mechanism (the breaking of supersymmetry, for example), this is essentially an act of faith.

The second age of yeast-dominance is now. This is the current era (beginning about five billion years ago) when dark energy (the yeast) took over from cold matter (the dough) and again is driving an accelerated exponential expansion of the world, but much more slowly than in the early inflationary epoch. These periods of vacuum-energy-dominated exponential expansion, as well as those of radiation

6.1 The Evidence for Dark Energy

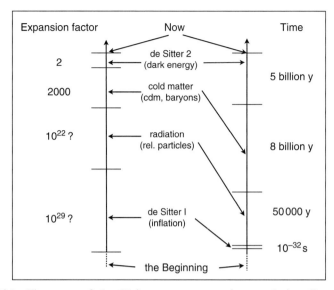

Figure 6.1. The ages of the Universe: an expansion- and time-line showing the dominant component at various eras. The left vertical arrow illustrates, approximately, the factor by which the Universe expands during each period. The length of the segments is very roughly proportional to the logarithm of the expansion factor. The right vertical arrow illustrates, schematically, the duration of each period, again logarithmic. The inflationary and radiation-dominated eras are indicated with a question mark because it is unknown when inflation begins and ends.

and matter dominance, are shown schematically in the time-line of Figure 6.1. Here we see that, while inflation was a short episode ($\approx 10^{-32}$ seconds), the Universe expanded by a large factor, possibly 10^{29}, during this period. The current de Sitter phase has lasted five billion years so far, but the Universe has only expanded by a factor of two.

The support for the initial extremely rapid expansion, the inflationary era, consists primarily in matters of principle (a natural explanation for the near-zero curvature and the observed isotropy of the Universe). But there is existent astronomical evidence for the present phase of exponential expansion – evidence leading to a determination of the contribution of this component to the composition of the world: the dark energy currently comprises about 70% of the total mass–energy density. But what does this actually mean? What is this substance, the slow-acting yeast, and what is the astronomical evidence? Operationally, it means that the Universe has, relatively recently, entered a de Sitter phase: the expansion law is exponential with a scale factor proportional to e^{Ht}, where t is cosmic time and H is a constant Hubble parameter. That is to say, the expansion of the Universe is accelerating. This is strange because ordinary radiation or cold pressureless matter cannot result in acceleration; the expansion can only decelerate due to the gravity of particles or photons. Even in a completely empty universe (the negative curvature model of Friedmann) the expansion rate is constant – no acceleration or deceleration.

The primary evidence for the accelerated expansion is the Hubble diagram, the apparent brightness vs. redshift, for type Ia supernovae (Figure 5.3), presumably objects with a standard maximum luminosity (after correction for the correlation with the decay time of the event). Any object of fixed luminosity, a standard candle, appears dimmer when it is further away, but type Ia supernovae appear to be even dimmer than expected in a non-accelerating universe – there is extra dimming by about 15% at a redshift near 0.5 that can be explained by the accelerated expansion. But how robust is this result?

Type Ia supernovae are thought to result from the explosion of a white dwarf (an evolved star that has exhausted its nuclear fuel) in a binary system with a normal star that is consuming hydrogen at its center – a dwarf and its companion locked in a spinning fatal embrace. Unlike the normal companion, the white dwarf is held up against its self-gravity by the pressure of degenerate electrons; which is to say, the electrons are so densely packed that their velocity, and therefore pressure, is due to the quantum exclusion principle rather than the usual thermal motion. And as shown by Subrahmanyan Chandrasekhar more than 80 years ago, such configurations are possible only if the mass of the dwarf is less than about 1.4 times that of the Sun; for a higher mass the velocity of the electrons approaches the speed of light and a degenerate star is unstable.[1] When the normal companion star evolves and becomes a red giant, matter flows from the giant to the dwarf and thereby increases its mass, which at some point exceeds that of the Chandrasekhar limit. The white dwarf collapses and then detonates in a nuclear explosion that completely disrupts the white dwarf and produces the observed supernova (this is the preferred model but there are alternatives).[2]

Locally, type I supernovae appear to exhibit a characteristic value for their peak luminosity; which is to say, the maximum power of the explosion is the same for all such events to within about 10%. At this peak luminosity of roughly 10 billion times the power of the Sun, the supernova power is comparable to that of its parent galaxy for about a week. These events occur in both general types of galaxies – spiral disk galaxies that contain a population of young as well as older stars – and spheroidal (elliptical) galaxies that are composed primarily of older stars. Of course there are also type II supernovae that are associated with young massive stars and are not standard candles; these occur only in spiral galaxies. The type I supernovae are easily distinguished from type II by their light curves (the manner in which the light varies with time) and by their spectra (there is an absence of the characteristic lines of hydrogen in type I supernovae). So all in all, type 1 supernovae would appear to be ideal cosmological probes.

But could there be systematic effects which produce the apparent extra dimming that we attribute to cosmology? Intergalactic dust would seem to be one possibility.[3] By obscuring background light, dust in the space between galaxies would also cause extra dimming for more distant objects. But now a number of supernovae at redshifts exceeding one have been observed, and the extra dimming is reversed – these more distant objects again appear to be brighter relative to the expectation for an accelerating universe. This is consistent with the transition from deceleration to acceleration (see Figure 5.1) or the emerging dominance of the dark energy over the cold matter at a redshift of less than one. This relative reversal of extra dimming at higher redshifts would be unexpected if the extra dimming were due to dust; the dust can only cause more dimming of more distant objects.

Another possibility would be an intrinsic variation of the properties of the actual supernova events with the age of the Universe (hence redshift). The proportion of young to old stars varies with cosmic time (more young stars at earlier epochs), so if the properties of the type I supernovae differ in young and old stellar populations this could lead to an intrinsic variation of the peak luminosity with redshift. It also might be expected that the abundance of heavy elements would increase with cosmic time (due to nucleosynthesis in stars), which would affect the opacity of the outer layers of the supernova progenitors as well as the composition of the fuel itself – fewer heavy elements in high-redshift objects. Perhaps such factors could combine to mimic the effects of cosmology on the Hubble diagram of supernovae. This would seem improbable and certainly more complicated than the cosmological explanation, but it would be useful to have other probes of the accelerated expansion; the result is, after all, of overwhelming significance for the deduced composition and the ultimate fate of the world. A better theoretical understanding of the astrophysics of the explosion event and its progenitors would add further confidence in this result.

There is a further alternative possibility: If we happen to live near the center of a large (several hundred million light years) under-dense region of the Universe, a local void, then it is possible to observe acceleration in the expansion of the relatively nearby Universe without any contribution from dark energy or a cosmological constant.[4] In a void, the expansion rate is higher because the counter-force of gravity is lower; this can mimic the effect of a cosmological constant while overall there is no acceleration of the universal expansion. There is, however, indirect evidence of the contribution of dark energy – evidence supplied again by the cosmic microwave background.[5]

The details of the CMB anisotropies are rather insensitive to dark energy. If the energy density of this medium is truly constant with the expansion of the Universe, as would be the case if it were represented by a cosmological term in Einstein's equation, then its contribution to the mass–energy density at decoupling (a redshift of 1000) would be one billion times smaller than that of the dark matter contribution. Dark energy could have no direct dynamical effect upon the acoustic oscillations as do the photons, the baryons or the dark matter. The argument in favor of dark energy comes from the deficiency in the density budget of these other components. The CMB acoustic peaks, primarily the standard meter stick provided by the first peak (the sound travel time horizon at decoupling), constitute convincing evidence that the Universe is flat and thus at a critical density ($\Omega_0 = 1$). But from the relative amplitudes of the higher-order peaks, there is also strong evidence that these other components add up to only 30% of the total energy density required for a flat universe ($\Omega_{\text{matter}} \approx 0.3$). So if 70% of the Universe is not in any of these familiar forms, then it must be in the form of dark energy, or so the argument goes. These proportions are consistent with the supernova results.

6.2 The Nature of Dark Energy

6.2.1 Zero-Point Energy

While we know what the slow-acting yeast does, we do not know what it actually is. What is this stuff that now dominates the energy density of the Universe – this strange fluid that maintains a more or less constant density as the Universe expands? It is simple, in fact trivial, to identify the dark energy with a constant term in Einstein's field equation, but this is really not an explanation. It is without justification and tells us nothing about the nature of this strange substance.

It would seem to be most natural to identify the dark energy with the zero-point energy of a quantum field, because this appears in Einstein's equation in the form of a cosmological constant. In the context of quantum mechanics, the vacuum is far from empty: there are virtual particles and their anti-particles, for example electrons and positrons, that constantly pop into and out of the vacuum. This does not violate conservation of mass–energy if these particles can exist only for a timescale that is consistent with the Heisenberg Uncertainty Principle, which is to say, for an interval of time less than $\Delta t \sim h/Mc^2$ (h is the well-known Planck constant and M is the mass of the virtual particle). For an electron–positron pair this would be about 10^{-20} seconds – a short but finite time.

For each quantum field (the electromagnetic field, for example) the energy density in these virtual particles can add up to be quite large: the total energy density is proportional to the rest energy of the virtual particle to the fourth power

$((Mc^2)^4/(h^3c^3))$; for the electron–positron pairs this would amount to a rest mass density of 60 grams per cubic centimeter, or about 6×10^{30} times larger than the critical density of the Universe! And this is only one quantum field. If we take the rest energy of the particle to be the Planck mass, about 2×10^{-5} grams (as might seem appropriate for a theory of quantum gravity), the zero-point energy would exceed the critical density by 120 orders of magnitude (10^{120}).

This is the traditional cosmological-constant problem. Identification of the dark energy with the zero-point energy of quantum fields leads to an impossibly large value.[6] It might be thought that some relief from this severe problem may be provided by the theory of supersymmetry, because fermion fields (corresponding to particles with half-integral spin) contribute a negative energy density while boson fields (particles with integral spin) contribute a positive energy density. Because in the context of this theory there is a partner boson corresponding to every standard-model fermion and vice versa, we might expect that for equal mass superparticles the contributions to the zero-point energy would exactly cancel. But the non-detection of superparticles so far means that they must be substantially more massive than the standard-model particles, so the cancelation is imperfect. Thus supersymmetry could reduce the vacuum energy density by about 60 orders of magnitude, which still leaves more than 60 powers of ten too much vacuum energy density.

All in all, the problem of the small magnitude of the cosmological constant must be considered unsolved. In the context of string theory (the so-called landscape version), the multiverse could provide an anthropic solution. Here, the number of possible vacuum states is large and we, of course, would inhabit a Universe with an appropriately small cosmological term. But this suggestion remains speculative and is not falsifiable.

6.2.2 Dynamic Dark Energy

The zero-point energy of a quantum field is truly like a cosmological constant – its energy density does not vary with time or position. But it is also possible that the vacuum energy density could be variable – that the energy density could vary with the expansion of the Universe, but differently than that of radiation or matter. This variable, or dynamic, vacuum energy is called *quintessence*,[7] following the ancient Greek description of the world as consisting of five elements – the familiar substances of earth, water, fire and air, but also a fifth essence that fills the celestial realm.

The rate at which the energy density of any fluid varies with expansion is determined by the equation of state that relates pressure p to density ρ; that is, $p = w\rho c^2$, where w is a number that depends upon the properties of the fluid. In an expanding universe, the dimensionless scale factor is that factor by which all distances increase with time; the volume increases as a^3. The general relationship between the density of mass–energy of a fluid and the scale factor is $\rho \approx a^{-3(1+w)}$. For pressureless matter (CDM or baryons), $w = 0$ and the density decreases inversely as the volume, $\rho \sim 1/a^3$. For electromagnetic radiation (photons) or other relativistic particles, $w = 1/3$ and the energy density decreases more rapidly with volume;[8] i.e., $\rho \approx 1/a^4$.

This same factor w also determines the effect of the fluid on the expansion of the Universe. For example, pressureless matter ($w = 0$) causes expansion to decelerate (the gravitation of the fluid acts counter to expansion). For any fluid that causes accelerated expansion, it must be the case that w is negative and less than $-1/3$ (the case of $w = -1/3$ drives linear expansion – no acceleration or deceleration, as in a negatively curved empty universe). For w between $-1/3$ and -1, the Universe accelerates but the energy density of the fluid does decrease, just more slowly than that of cold matter or radiation (for $w = -1$, the cosmological constant case $\rho \approx$ constant; the density does not decrease at all).[9] For an accelerating expansion, it must be the case that the effective pressure of the dominant component is negative; it is the negative pressure that drives the acceleration.

When w is negative but not -1, it is possible to have a dynamic dark energy – a medium that drives acceleration but with a slowly decreasing energy density. Such a dynamic dark energy can be realized if the dark energy is a potential energy associated with a cosmic field, as in inflation. This is to say, the evolution of vacuum energy density in the Universe is like a ball rolling down a potential hill or well (as in Figure 6.2 or 6.3), and the dark energy density at any time corresponds to the height of the ball above the absolute zero point at that epoch. In the general case, w is not constant but varies with cosmic time.

The way in which w and therefore the density of the dark energy varies with expansion or cosmic time depends upon the shape of the potential well. For example, if the slope of the well is steep initially, but becomes more gradual as cosmic time proceeds, then the model can be described as "freezing"; the dark energy is initially quintessence, but then the rolling of the ball slows down and becomes, in effect, frozen as a cosmological constant (a potential of the form $V(\phi) \approx \phi^{-2}$ could meet this condition; see Figure 6.2). If, on the other hand, the slope is not so steep at early times (as, for example, the potential has a quadratic form $V(\phi) \approx \phi^2$), initially the potential is frozen as for a cosmological constant, but then it "thaws," becoming quintessence later. It is even possible to produce a model in which initially the field is frozen, then thaws and oscillates; such oscillations would constitute cold dark matter in the form of long-wavelength bosons. Slowly the oscillations damp and the Universe settles into a minimum which, if greater than zero, behaves as a cosmological constant (position 3 in

6.2 The Nature of Dark Energy

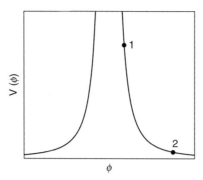

Figure 6.2. Freezing: This potential $V(\phi) = 1/\phi^2$ can lead to a dynamic dark energy associated with the scalar field ϕ that, near the instant of the Big Bang (at position 1), rapidly decreases as a function of cosmic time. But then, near the present epoch, when the slope of the potential becomes much smaller, the dark energy freezes near a fixed value – effectively as a cosmological constant.

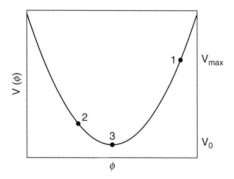

Figure 6.3. Thawing: This plot demonstrates a possible scalar field-model of dynamic dark energy and cold dark matter. It shows the self-interaction energy $(V(\phi))$ of a hypothetical scalar-field (ϕ) having a quadratic dependence of energy density on the scalar-field strength. When the timescale for the scalar field to roll is longer than the age of the Universe, the field is frozen at position 1 and exhibits a large energy density (V_{max}). When the age of the Universe exceeds the rolling timescale, the field "thaws" and rolls down the potential well past the minimum (position 2) and then begins to oscillate about the minimum at position 3. The oscillations would constitute cold dark matter, but they are slowly damped as the Universe expands (this damping corresponds to the dilution of the dark matter with the expansion of the Universe). The field settles toward the lowest point – the vacuum at position 3. This would constitute an effective cosmological constant (V_0).

Figure 6.3). This could provide a large cosmological constant, followed by matter domination, followed by a smaller cosmological constant that finally dominates, as we seem to observe.

There are aesthetic and empirical problems with scalar fields as a mechanism for dark energy. The scalar field does not solve the problem of the small cosmological constant; it is simply assumed that the minimum of the potential energy is very small. Moreover, the mass scale of the scalar field (which determines the slope of the potential curve and allows for slow rolling or freezing) is very much lower than that of other mass scales in particle physics. For example, the mass associated with the newly discovered Higgs particle is 140 GeV, whereas that associated with the hypothetical scalar field providing the dark energy has to be 44 orders of magnitude less; such a large hierarchy of scales seems quite unnatural. An additional problem concerns the possible coupling of the scalar field to matter: the coupling to matter must be considerably weaker than that of gravity; otherwise, we could detect the scalar as a "fifth force" on particles, leading to a possible violation of the Equivalence Principle (the independence of gravitational acceleration on mass or composition of the particle – Galileo's Tower of Pisa experiment). Fifth forces are constrained locally to be very small indeed,[10] so scalar fields as an explanation of evolving dark energy – quintessence – remain speculative.

It is also possible that the accelerated expansion could result from a modification of the Einstein field equation – modified gravity – that is significant over cosmological distances and timescales.[11] However, given the consistency of the standard cosmology at early times, or high redshifts ranging from the epoch of primordial nucleosynthesis to that of decoupling of photons and matter, such modifications should become effective primarily at low redshift or recent epochs.

There have been a number of suggestions for modified gravity, but these are all speculative.[12] Such ideas can be constrained by measuring the evolution of the equation-of-state parameter, w, over cosmic time. The acoustic oscillations probed by the CMB anisotropies are consistent with $w = -1$, the case of a cosmological constant, but are not definitive for the problem of evolution. This is because the CMB provides a snapshot of the Universe at a single epoch, and one in which the vacuum energy has a negligible dynamical effect. Combining the CMB observations with the baryon acoustic oscillations, as observed in the large-scale distribution of galaxies, can be more definitive. The baryon acoustic oscillations are particularly useful in this respect because this phenomenon allows a determination of the evolution of the Hubble parameter, $H(z)$, at a much lower redshift than that of the CMB (typically $z \sim 0.35$ for the galaxies in the Sloan Digital Sky Survey). So far, the results remain consistent with $w = -1$, but the improving data should permit more precise constraints.

A complementary approach in probing the evolution of dark energy involves the use of gravitational lenses. This phenomenon, predicted long ago by Einstein, requires the near alignment of a bright background source, such as a galaxy nucleus (a quasi-stellar object) or an entire galaxy, with a massive foreground object – the lens, typically a galaxy or a cluster of galaxies. The image of the background source is altered by the gravitational field of the foreground object in a way that depends upon the mass and mass distribution in the lens, the degree of alignment and the cosmology via the distances of the lens and the source, in particular the angular-size distance – the distance that we would determine by measuring the angle subtended by a standard meter stick. The lensing can be strong, with the formation of multiple images of the background source requiring close alignment, or weak, with the slight distortion of the source image arising in approximate alignment. Thus, in principle, by observing lenses at different redshifts it should be possible to measure the evolution of dark-energy density with redshift, provided that the mass distribution in the lens can be accurately modeled. In a number of cases this appears to be possible for both strong and weak lenses, although there remains a problem of degeneracy with the mass distributions.[13]

6.3 Dark Energy and Fundamental Physics

We have considered several possibilities. The dark energy could be simply a constant in Einstein's equation, but this has no evident basis in physics. It could be the zero-point energy of a quantum field – an idea that is well-motivated. The difficulty is that the estimated value of energy density is many orders of magnitude too large; the Universe would have been dominated by vacuum energy from very early epochs and matter would never have condensed into the familiar structures. The vacuum energy could be dynamic and associated with the potential – the self-interaction energy – of a long-range scalar field. This leads to many possible scenarios for the evolution of the dark energy, and the scalar field could reveal itself by so far unseen long-range forces. But again, there is no connection with known physics, and the scalar field potential does not solve the problem of the large cosmological constant resulting from zero-energy quantum fields. In the same vein, the accelerated expansion could be connected with new gravitational physics beyond Einstein's theory, but there is not yet a convincing model.

It is evident that the nature of dark energy remains a mystery, and one that connects with the most fundamental problems of modern physics. This is like trying to bake our cake without any idea of the current dominant ingredient – the slow-acting yeast that causes the present exponential expansion. Without more understanding of its nature it is difficult to formulate the relevant questions, let alone provide the answers. However, several issues come to mind: Is there a relation between the present de Sitter phase – the exponential expansion apparently observed in the Hubble law of distant supernovae – and the hypothetical initial expansion associated with early dominance of vacuum energy – the process that we call inflation? Will the present dark energy disappear – will it thaw and approach a true zero-energy vacuum state? Or will it freeze, as when the scalar field initially drops rapidly but then hangs up at a finite value (as in a ϕ^{-2} potential)? Will the present acceleration continue forever, with all external galaxies eventually disappearing beyond the cosmological horizon, or will the Universe eventually re-collapse as the dark energy decays to zero? Will the Universe undergo a big rip (with $w < -1$) that will ultimately tear apart the very atoms? Because we do not understand the nature of dark energy, we cannot yet claim to understand the world and its ultimate fate.

Dark energy is the yeast in the cosmic cake. We know what it does, but not what it is. Because the evidence for dark energy is and very probably can only be astronomical, can we ever understand its "true nature"? Is an operational description all that is possible? Have we reached an impenetrable boundary of science? A related question, posed several years ago by Simon White, is whether or not dark energy is a useful subject for astronomical research or a waste of time and resources.[14]

I take an optimistic view of this issue. A more complete understanding of dark energy can certainly be achieved by constraining the evolution of its energy density, or more simply, the equation-of-state parameter w. We know that this can be done via more complete and precise observations of the large-scale distribution of galaxies and thus the baryon acoustic oscillations in two dimensions – perpendicular and along the line of sight. Gravitational lenses, both strong and weak, also probe the cosmological distance – the angular-size distance that depends on the relative contribution of dark energy at different redshifts. So there is hope at least for an empirical description of the evolution of the dark-energy density.

In the realm of theory, ideas at present are speculative (scalar-field models, modifications of general relativity), but this is the beginning of the process of developing actual models. It would be of interest to consider hypotheses in which the dark energy is related to other phenomena, such as the dark matter problem and the near coincidence of the energy densities of these two fluids at the present epoch. The cosmic acceleration introduced by dark energy ($\sim cH_0$) also makes an appearance with respect to the phenomena of local systems (e.g., galaxies), but this is a topic for Chapter 8.

7
Dark Matter

If dark energy is the yeast in the cosmic cake, the dark matter is the flour – it provides the substance and the material content. At first thought, it would seem to be more substantial and comprehensible than dark energy, so we might expect to have a better understanding of this medium. As a concept it has been around longer and is easier to visualize: we can all imagine billiard balls bouncing around in the gravitational potential well of a galaxy.

In fact, dark matter is stranger than the billiard-ball picture. The particles that are thought to comprise dark matter are non-standard – not the usual protons and neutrons of baryonic matter. They are electrically neutral, they do not interact directly with photons, and they interact rarely with baryons and rarely with themselves; it is as though the billiard balls can pass right through each other and everything else. They are notoriously difficult to detect by any direct non-astronomical technique, which is to say, the matter is very dark indeed. The essential interaction with normal baryonic matter and photons is gravitational, and the direct long-range gravitational influence of dark matter upon observed systems – clusters of galaxies, spiral and elliptical galaxies, dwarf spheroidal galaxies – is the primary evidence for its existence. So, in fact, this medium is just as peculiar and poorly understood as dark energy, and its existence requires new and exotic physics.

7.1 Evidence for Dark Matter in Galaxies and Galaxy Systems

The very first evidence of the dynamical effects of a substantial unseen component in a bound gravitational system was discovered in 1933 by Fritz Zwicky, not in individual galaxies but in a cluster of galaxies. Here the individual galaxies are moving much too fast if the only mass is in the visible starlight of the galaxies. The system should be unbound – the galaxies should not be seen to be clustered but flying away from each other, unless there is an unseen component holding the cluster together. Zwicky first used the term "dark matter" to describe this unseen component and estimated that it must outweigh the visible matter by a factor of several hundred. Subsequently, much of the "missing mass" was found to be in the form of X-ray-emitting hot gas (baryons), which can outweigh the visible stars in galaxies by a significant factor. Still, the true dark matter must be six or seven times more massive than the detectable component.[1]

The dark mass is also detected via gravitational lensing (assuming the validity of general relativity on cosmic scales). As described in Chapter 6, the light from a distant source – a galaxy or galactic nucleus – is deflected by the gravitational field of an intervening mass concentration, such as a cluster of galaxies or an individual galaxy. The astronomer observes the resulting distortion of the images of background objects – weak lensing – or, in more extreme cases, the formation of multiple images or rings – strong lensing. A beautiful example of a cluster gravitational lens is shown by the cluster A2218 (Figure 7.1) that exhibits both weak and strong lensing.[2] The detailed form of the distortions or multiple images

Figure 7.1. Abel 2218 is a beautiful example of a cluster of galaxies acting as a gravitational lens. This cluster exhibits both the phenomena of strong lensing – arcs and multiple images of background galaxies primarily about individual cluster galaxies – and weak lensing – the distortion of background images by the cluster as a whole. When interpreted in terms of general relativity, the lensing mass of the cluster is comparable to that estimated by the temperature and distribution of hot gas. Image courtesy of NASA.

depends upon the degree of alignment and cosmological distances to the source and the lens, but primarily upon the total mass and its distribution in the lens.³ The results, interpreted in the context of general relativity, generally agree with those provided by more traditional methods (rotation curves, internal kinematics of clusters) regarding the existence, total mass and distribution of dark matter. This is significant because the traditional methods use kinematics of slow-moving, non-relativistic particles and require only the use of Newtonian dynamics, but lensing involves the effect of gravity on relativistic particles (photons) and requires a relativistic theory of gravity.⁴

The most precise description of the distribution of dark matter in bound gravitational systems is provided by the observed rotation curves of spiral galaxies. A rotation curve is a plot of the rotational velocity in an astronomical object against distance from the center of the object. The very first measured rotation curve of an astronomical system was that of the Solar System; here the rotational velocity about the Sun, as evidenced by the planets, falls as the inverse square root of the distance from the Sun ($V \propto 1/\sqrt{R}$); this is Kepler's law for circular orbits and the rotation curve is said to be Keplerian. The rotation curve of the Solar System is consistent with all of the mass in the system being located at the center (the Sun) and with Newton's universal law of gravity (this was a primary motivation for the inverse-square form of Newton's law).

In galaxies well beyond the visible mass, the rotation law differs dramatically from Keplerian form – the rotation velocity at large distances from the center of the object is constant with the radius, as we see in Figure 7.2. This is the

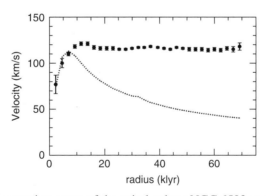

Figure 7.2. The rotation curve of the spiral galaxy NGC 6503 as observed in the 21-cm line of neutral hydrogen. The points with error bars give the rotational velocity (km/s) as a function of the distance from the center of the galaxy in kilo-light years (thousands of light years). The dashed curve is the rotational velocity predicted if the mass distribution is traced by the light intensity with a mass-to-light ratio of one in solar units. The discrepancy between the curve and the points becomes very apparent in the outer regions.

rotation curve of a nearby spiral galaxy, NGC 6503, as measured in the 21-cm line of neutral hydrogen that extends far beyond the luminous stellar component of the galaxy.[5] The points with error bars show the measured rotation velocity plotted against distance from the center of the galaxy in units of thousands of light years. The dashed curve is the rotation velocity that is calculated via Newtonian gravity from the observed distribution of visible light and detectable gas, assuming that the stellar population has a constant mass-to-light ratio of one in solar units (a reasonable assumption). The difference between the points and the curve dramatically demonstrates the discrepancy.

It is evident here that the discrepancy is most apparent in the dark outer regions where the rotation curve should be falling in an almost Keplerian form but remains constant. As for the Solar System, the rotation curve extends into the region where no light is found, but unlike that of the Solar System, the rotational velocity does not decline. In the context of Newtonian dynamics, this can only be understood if the mass distribution extends well beyond the visible starlight – which is to say, the mass in the outer regions is dark.

The phenomenon is general.[6] It is true in virtually all disk galaxies that have been observed in sufficient detail that the rotation curve remains flat beyond the visible disk. The constancy of the rotation velocity out to large distances from the parent galaxy appears to be as regular as a law of nature. In the context of the dark matter hypothesis, this implies that every visible galaxy is embedded within a more extensive dark component, a halo of dark matter, having a very specific density distribution: the density must fall as $1/r^2$.

We might think that, if dark matter cannot be seen, then it can be distributed in any way necessary to match the observations of galaxy rotation curves; which is to say that the theory has no predictive power. In fact, cosmic N-body simulations involving millions of particles interacting only gravitationally in an expanding universe have provided a detailed description of the structure of halos on all scales – from sub-galactic objects to the great clusters of galaxies. The density distribution in these structures has a universal form characterized by a length scale, say R_s, and a central density, ρ_0, both of which determine a velocity scale, v_0. At radii less than R_s the density varies as $1/r$; i.e., there appears to be an inner cusp – the density becomes very large as the radius approaches zero. At radii larger than R_s, the density decreases as $1/r^3$, which should lead to an asymptotically decreasing rotation velocity (albeit slowly as $\sqrt{ln(r)/r}$). This is the famous Navarro–Frenk–White (NFW) halo.[7] A declining asymptotic velocity is, at first sight, in disagreement with the observed asymptotic form of galaxy rotation curves ($V \sim$ constant), and constitutes a *prima facie* falsification of CDM as it is perceived to be. In practice, observed rotation curves can be reproduced with such a halo model in combination with the observed baryonic matter by a careful matching of R_s and ρ_0 with the disk length scale and rotation velocity, but then this becomes an exercise in parameter fitting and not predictive science.

With respect to galaxy phenomenology, there are several problems with the standard form of CDM halos. The predicted inner cusp is often not consistent with the observed rotation curves; the observations are more generally consistent with a halo having a constant density core.[8] The cosmic N-body calculations predict the presence of many sub-halos – many more than are actually observed in the neighborhood of spiral galaxies such as the Milky Way.[9] The unobserved satellites are therefore assumed to be void of baryonic matter (possibly blown out by winds from first-generation stars) and are therefore totally dark. These problems (core–cusp and the missing satellites) are the so-called small-scale problems with CDM, but there are even more fundamental issues (to be discussed in the next chapter).

7.2 Cosmological Evidence for Dark Matter

Most of the cosmological evidence has been described in Chapters 4 and 5 and is reviewed here. Before the discovery of dark energy via the observation of the present accelerated expansion of the Universe, dark matter appeared to be required in order to make up the density budget of an $\Omega = 1$ universe as required in the context of the inflationary scenario (as in SCDM). The standard theory of nucleosynthesis, combined with measurements of the primordial abundance of light elements, leads to the result that the baryons can only supply a few percent of this closure density; the remainder must be due to non-relativistic dark matter particles that do not participate in the process of nucleosynthesis.

In addition to this global consideration, cold dark matter on a cosmological scale is required for the formation of structure in the context of gravitational collapse (with standard gravity) in an expanding universe of finite age. The problem is that positive density fluctuations in the baryonic fluid can only begin to collapse when the protons and electrons combine to form neutral hydrogen and thereby decouple from the photons at a redshift of about 1000. But then the density fluctuations that form structure can grow by at most a factor of 1000 up to the present epoch. This means that the initial amplitude of the fluctuations must have been at least 1/1000 at the epoch of decoupling, which would lead to temperature fluctuations in the background radiation of the same order of magnitude. Such large fluctuations are clearly not present in the CMB. The problem can be overcome by the presence of dark matter that does not couple directly with photons. Fluctuations in this dark component can begin collapsing well before the decoupling of baryons and photons; then later on after decoupling the baryons fall into the pre-existing gravitational potential wells of the dark component.

The detailed observations of fluctuations in the CMB on angular scales of less than one degree provide quantitative evidence for this cosmic dark matter. The amplitudes of the acoustic peaks in the power spectrum appear to require the existence of rigid, pre-existing potential wells in which the oscillations of the baryon–photon fluid take place. In detail, this sets the contribution of dark matter to the current density budget of the Universe to be about $\Omega_{cdm} = 0.25$; with baryons, the total matter contribution to the present density of the world is about 30% of the closure density, the remainder comprising dark energy.

The Hubble diagram of distant supernovae agrees with this division between matter (30%) and dark energy (70%). Moreover, X-ray observations of the hot gas in clusters of galaxies, providing estimates of the mass of hot gas as well as the total dynamical mass, also apparently sample these same cosmic contributions of baryonic and dark matter at a ratio of one to six.

So, on a cosmological scale, the picture appears to be consistent and the evidence for dark matter is most convincing. The composition of the world is about 25% dark matter that promotes the early formation of structure; the matter is cold and that allows it to cluster on small scales – certainly down to the scale of galaxies. This is where the paradigm impinges upon well-studied local systems and where difficulties arise. We should also keep in mind that the evidence for CDM, cosmological as well as local, is gravitational. The existence of dark matter with its perceived properties requires the assumption of the validity of general relativity (or Newtonian dynamics) from galactic to cosmic scale.

7.3 The Nature of Dark Matter

Certainly the cosmic dark matter cannot consist of ordinary protons and neutrons – baryons – because there are not enough of them. Both from considerations of primordial nucleosynthesis and from the direct observations of the CMB anisotropies, we know that baryons comprise only 5% of the mass density of the present Universe. And even then, roughly one half of these baryons expected to be present are not observed – at least, not in the present low-redshift Universe.

The shortage is severe in galaxies. From weak gravitational lensing it is evident that the dark halos extend to between 300 and 600 thousand light years – roughly 10 or 20 times the radius of visible galaxies.[10] This means that the dynamical mass of galaxies keeps growing linearly with radius to ten or twenty times the visible or baryonic mass. But the universal ratio of dark matter to baryons is only six to one, not twenty to one. The deficiency in baryons is even more dramatic in dwarf or low-mass galaxies. So where have all the baryons gone? Why is the problem of missing baryons so much more severe in galaxies than in clusters of

7.3 The Nature of Dark Matter

galaxies, where the ratio of dark to visible mass is more like the universal ratio? What has happened to the baryonic content of galaxies?

The current view is that these baryons have been blown out of galaxies by various processes connected with stars and their evolution – processes such as winds from young stars and supernova explosions from evolved stars, or by the high luminosity of massive black holes in active galactic nuclei. Moreover, there are additional hydrodynamical mechanisms that can remove gas from galaxies but not remove the stars or the supposedly dissipationless dark matter: stripping of the gas in dense clusters of galaxies due to the motion of galaxies through the gaseous intracluster medium (ram pressure), or collisions between protogalaxies (or protoclusters), in which the dissipationless components pass through one another leaving the gas behind (as in the "bullet" cluster).[11]

All of these more-or-less random processes would be expected to remove a greater fraction of the gas in dwarf galaxies, where it is more loosely bound, than in massive galaxies. But where are the baryons now? Why is it that roughly half of the baryons known to be present are not observed? The missing baryons are thought to reside in a warm intergalactic medium: a tenuous diffuse gas between the galaxies at a temperature of 100 000 to 10 million degrees. Such gas would be difficult to detect because it would be mostly ionized (no spectral lines) and would emit primarily low-intensity continuum radiation in the far-ultraviolet or low-energy X-rays.

This is, however, a comparatively trivial issue. The real mystery concerns the identity of the true dark matter – the 25% of the Universe for which there is direct dynamical evidence, but which is not detected through the emission of any radiation or energetic particles. What is this mysterious stuff that promotes structure formation and comprises the pre-existing potential wells in which the baryon–photon fluid oscillates? What is the dominant dark component of galaxies and clusters of galaxies that is implied to be present by their internal kinematics or gravitational lensing?

The requirement that the dark matter should be cold – i.e., non-relativistic when it stops interacting with photons (decouples) – rules out particles such as standard neutrinos, that have a mass of less than a few electron volts but stop interacting with photons when their kinetic energy is one or two million electron volts. These particles are highly relativistic when they decouple and, via free-streaming at the speed of light, smooth all fluctuations on a small scale (such as that of galaxies). But if a dark matter particle has, for example, a mass of 100 GeV (one hundred billion electron volts) but stops interacting when the kinetic energy of particles and photons has fallen below this value, then it is obviously non-relativistic at that epoch. Such dark matter is free to collapse and form structure down to the scale of dwarf galaxies (or smaller).

There have been numerous suggestions for the nature of dark matter particles, but the most well-motivated proposal, from the point of view of theoretical physics, is that of WIMPs – "weakly interacting massive particles." These are hypothetical particles that are stable (long-lived), electrically neutral, and weakly interacting with themselves and with baryons and photons (otherwise they would have been directly detected). They are "thermal relics" – particles that, early on, when the Universe had existed for less than 10^{-10} seconds, were in thermal equilibrium with photons. This means that photons and WIMPs have the same temperature: they can convert back and forth, one into another, via pair creation and annihilation. The WIMPs are also relativistic, having the same kinetic energy and the same number density as the photons. When the Universe expands and cools to the point where the energy of photons is less than that of the WIMP rest-mass energy (at an age greater than 10^{-10} seconds for a 100 GeV WIMP), there is no longer pair creation by photons and the WIMPs begin to annihilate (they are their own anti-particles) until the density is so low that they can no longer find each other in the expanding universe. They are clearly non-relativistic at this point; which is to say, they are cold and able to promote the formation of structure on all scales.

There are no such particles within the standard model of particle physics, so they must arise in the context of theories that go beyond the standard model – theories such as supersymmetry. As described above (Chapter 5), supersymmetry effectively doubles the number of particle species by providing a superpartner for every standard model particle. None of these hypothetical new particles is yet detected (as of December 2015) in high-energy particle accelerators (such as the LHC), so they must be massive and unstable. Only one of these superparticles is likely to be stable – that superpartner with the lowest mass,[12] because there is nothing for it to decay into. The cosmic abundance of WIMPs is fixed by the cross section for WIMP–anti-WIMP annihilation – basically this determines the number of WIMPs left over after decoupling from photons and subsequent annihilation. If that cross section is like that of the weak interaction (that is to say, on the order of $\sigma_a \approx 10^{-36}$ cm^2), then it can be shown that the present density of WIMPs is comparable to closure density, i.e., $\Omega_{\text{WIMP}} \approx 1$ (this has been called the "WIMP miracle").[13] Smaller annihilation cross sections would leave too many WIMPs and larger cross sections not enough. Because the cross section is reasonable for such particles, they are said to be the most well-motivated dark matter candidate, although, as of writing this there is no independent evidence for their existence nor for the validity of supersymmetry.

Even if supersymmetry is right and such particles do exist, they are not necessarily sufficiently abundant to comprise the dark matter. However, the fact that they annihilate each other and interact weakly with normal baryons raises the possibilities of indirect and direct detection – possibilities that are now being aggressively pursued. If they were detected by their occasional non-gravitational interactions, it would be one of the most significant scientific discoveries of all time and an independent confirmation of the ΛCDM paradigm. This is generally appreciated by the scientific communities involved and has led to major projects to find the dark matter by its non-gravitational effects; this effort has interesting sociological as well as scientific aspects.

7.4 The Science of Dark Matter Detection

There are two broad strategies for the non-gravitational detection of dark matter: indirect and direct detection. Indirect detection is the identification of signals in the sky that can emerge from dark matter annihilation, such as an otherwise unexplained source of gamma rays or high-energy neutrinos, or an unexpected abundance of anti-particles such as positrons in the cosmic rays that constantly bombard the Earth. Direct detection is the identification of dark matter particles by their interactions with baryonic matter in terrestrial experiments. Of these two methods, direct detection provides the most unambiguous evidence for the existence of dark matter particles. This is because, with indirect methods, there are most often competing astrophysical effects that can be confused with a dark matter signal. Here I first consider indirect methods, because of recent alleged positive detections.

7.4.1 Indirect Detection of Dark Matter

WIMPs, as neutral particles, are very possibly their own anti-particles; therefore when the density is high, they can interact among themselves and annihilate, producing various annihilation products such as standard-model energetic particles, γ rays and neutrinos. This would happen in the early Universe after the WIMPs have decoupled from photons and the density is still high. Then, as the Universe expands and the density declines, the WIMPs gradually stop annihilating, leaving the present cosmological relic. But later, when structure forms, fluctuations in the WIMP fluid collapse, the WIMP density increases and again the WIMPs begin to slowly annihilate. So the idea is to look at regions of high dark matter density, such as galactic nuclei and dwarf galaxies and find the signals of WIMP annihilation. A "smoking gun" signature of such a process would be a sudden decline in the intensity of radiation above the rest energy of the dark matter particle – in effect, a spectral line in high-energy emission.

The Fermi Gamma-ray Space Telescope is an instrument with the potential to detect gamma rays from environments where the density of WIMPs is expected to be high, such as dwarf spheroidal galaxies. In a recent analysis of six years of data from 15 dwarf galaxies near the Milky Way, no signal was detected, and the upper limit is significant: the annihilation must be less than 10^{-36} cm^2 for particles of mass lower than 100 GeV.[14] This, as mentioned above, is the critical value of the cross section for production of the cosmic abundance of WIMPs, so this constraint would appear to close a significant window on WIMPs of lower mass.

However, the Fermi telescope does detect an extended source near the center of the Milky Way Galaxy. Numerical simulations of the formation of galaxy halos from CDM predict that the halo has a centrally peaked density distribution increasing as $1/r$ into the very center. If the density of dark matter in the halo of the Milky Way is of the NFW form, then the cusp-like density ($1/r$) at the center should produce the brightest dark matter annihilation signal in the sky, so it is of interest that a source is observed in this direction. But there are complications. The Galactic Center is a complex region containing a massive black hole at the very center ($\approx 4 \times 10^6$ solar masses), a number of newly formed stars within several light years of the center, and an extended star cluster. There are many possible contaminating sources of gamma rays. Lisa Goodenough and Dan Hooper[15] have argued that the spectrum of gamma rays is consistent with the annihilation of dark matter particles, having a relatively low mass of 28 GeV and an annihilation cross section comparable to that required for the WIMP miracle. But then this would appear to be inconsistent with the limit for dwarf galaxies. An alternative explanation for this extended gamma-ray emission is a population of very short-period pulsars (neutron star remnants of supernovae), which are distributed similarly to the inner stellar cluster.

This illustrates how difficult it is to unambiguously detect dark matter by these indirect astronomical means.[16] There is almost always a host of competing astrophysical sources that provide a contaminating background.

Another recent example of this is the unexpectedly large number of positrons found locally (in near-Earth orbit) by the Alpha Magnetic Spectrometer (AMS) on board the International Space Station.[17] This instrument is designed to detect the anti-matter component of energetic cosmic rays, and in 2013 Samuel Ting, the principal investigator, reported that there was a significant, and unexpected, increase in the positron-to-electron ratio at energies of 10–250 GeV. There was no variation of this ratio with time or with direction.

Cosmic rays are relativistic particles – atomic nuclei, electrons and other species – that continually bombard the Earth from galactic and even extragalactic space. Their existence and nature have been understood for a century: the particles detected at the surface of the Earth are secondary energetic particles produced by collisions of primary cosmic rays arising in the Galaxy at large with atomic nuclei in the atmosphere of the Earth. To determine the nature of these extraterrestrial

7.4 The Science of Dark Matter Detection

cosmic rays – their composition and energy distribution – it is necessary to go above the atmosphere of the Earth. This is the advantage of instruments such as the AMS that can measure the anti-matter content of the primary cosmic rays – the electrons and positrons, protons and anti-protons, and heavier atomic nuclei – as a function of their energy.

The main source of cosmic rays is thought be galactic supernovae that accelerate particles up to relativistic energies in the shock waves produced by the explosion in the surrounding interstellar medium. These events inject primarily matter rather than anti-matter into the cosmic ray population, and, in the case of the AMS measurements, that means electrons rather than positrons. However, the high-energy cosmic rays collide with the ambient interstellar medium and produce secondaries (i.e., secondaries are not only produced in the atmosphere of the Earth but also in the intervening space), and these secondaries will contain some fraction of positrons. This appears to account for most of those positrons observed at low energies.

An increase in the fraction of positrons with increasing energy of cosmic rays could be a sign of dark matter annihilation – for example, with dark matter particles annihilating directly into positrons and electrons. Again, a definite signal would be a sudden decrease in the positron fraction below the rest energy of the dark matter particle. There are, however, possible astrophysical backgrounds, for example pulsars – rapidly spinning magnetized neutron stars that generate strong electric fields near the surface. These electric fields accelerate electrons and positrons into a wind that surrounds the surface but these particles are subsequently released impulsively through the disruption of the wind. To determine the energy distribution of positrons and electrons locally, the astrophysicist has to calculate how these particles produced by different sources diffuse throughout the Galaxy, where various processes (re-acceleration, energy loss, convection) can affect their spectrum, and finally how their energy is modulated near the Sun (the heliosphere) by the solar wind and magnetic field.[18] It is a complicated problem with a number of uncertainties. Nonetheless, the standard model of astrophysical sources, overall, is consistent with the observed ratio of positrons to electrons and its dependence upon energy. Certainly the "anomalous" ratio of positrons to electrons so far presents no compelling evidence for dark matter annihilations in the Galaxy.

Another possibility for indirect detection of dark matter particles involves the observation of neutrinos. Neutrinos are very light, neutral particles that interact infrequently with baryons and that were first noticed in the radioactive decay of unstable nuclei. They may be produced directly in the annihilation or decay of a dark matter particle and then detected in a neutrino telescope on Earth. There are now several such devices, with the largest and most sensitive buried beneath the ice near the South Pole – IceCube. This neutrino telescope consists of thousands of light-sensitive detectors (photo-multiplier tubes) distributed throughout a cubic kilometer of ice. In very rare interactions of neutrinos with atomic nuclei in the ice, a muon can be produced – a charged unstable particle that travels hundreds of meters in the ice before decaying. The muon is highly relativistic and therefore during its trajectory it can emit Cherenkov radiation – radiation that is emitted in a narrow cone about the velocity vector of the particle because the muon is traveling faster than the speed of light in the medium of the ice. The radiation is then detected by a number of the photomultipliers along its trajectory, thus making it possible to determine the path and energy of the muon. The path and energy are that of the original neutrino, so the enormous array provides data on the distribution of neutrinos arriving from the atmosphere or from space. The background is primarily due to muons created in cosmic ray interactions in the upper atmosphere, so an obvious way of discriminating against these events is to look only at upward moving muons – those arriving from the opposite side of the Earth – because such cosmic ray muons could never pass through the Earth, unlike neutrinos.

If a dark matter particle were to decay directly into neutrinos, it would be expected to produce a mono-energetic distribution of neutrinos – a spectral line in other words – at very high energy. So far, the only extra-terrestrial neutrinos definitely observed are at extremely high energies – from 100 000 to 1 000 000 GeV, or 0.1 to 1 PeV.[19] These neutrinos have an energy distribution that is consistent with a power law (roughly proportional to E^{-2}), as do the cosmic rays in this energy regime, so they are probably produced by cosmic ray interactions with ordinary matter or photons in energetic sources of cosmic rays – sources such as active galactic nuclei.

So there is no evidence for the detection of dark matter annihilation directly into neutrinos, but there is another possibility for dark matter detection via neutrinos. A nearby massive astronomical object, the Sun (or the Earth) for example, can trap dark matter particles in its interior. This happens because occasionally a dark matter particle on an orbit that carries it into the solar interior will interact with an atomic nucleus in the solar interior and lose energy. It will thus be trapped in the solar interior. In this way a concentration of WIMPs can build up in the solar interior. The density of WIMPs trapped by this process will keep increasing until the decay rate is equal to the capture rate; there will thus be a steady state distribution of WIMPs within the Sun and a steady annihilation rate. The only products of this annihilation that can make it directly to the solar surface and into the surrounding space are neutrinos, so the idea is to look for high-energy neutrinos coming directly from the Sun – a perfect objective for a neutrino telescope like IceCube.[20]

In a steady state, the rate at which neutrinos are emitted by the Sun is equal to the rate at which WIMPs are captured. Therefore the detection rate by IceCube depends only upon the WIMP–nucleon cross section and not directly upon the WIMP–WIMP annihilation rate. The non-detection directly constrains the process of interaction of WIMPs with nucleons and not the process of annihilation. An additional caveat is that the nuclei with which the WIMPs interact are those of the most abundant element – hydrogen (protons). This makes the rate of interaction quite sensitive to whether or not there is a dependence upon spin of the target nuclei (the net spin of protons is 1/2, larger than that of heavier nuclei, such as xenon, used in direct detection experiments). So the most sensitive constraint will be upon spin-dependent interactions.

It turns out that IceCube has so far detected no neutrinos from the annihilation of WIMPs trapped in the Sun, and the constraints on spin-dependent interactions are competitive with direct detection experiments. In general, it is safe to say that there is no convincing evidence for dark matter from indirect detection experiments – in spite of the optimism of many of those involved.

7.4.2 Direct Detection of Dark Matter

In the context of the standard cosmological paradigm, the visible disk of the Milky Way Galaxy is surrounded by a dark halo composed of the same matter that comprises the dominant material component of the Universe. This halo is supported against its own gravity primarily by the random motions of the component dark matter particles having a velocity spread on the order of 200 km/s. In the standard halo model of the galaxy, the local density of dark matter is 0.35 GeV/cm^3 or 6×10^{-25} g/cm^3. For a WIMP mass of 100 GeV, this would amount to a particle density of 3500 WIMPs per cubic meter. Taking a random velocity of WIMPs to be that typical of the Milky Way, the flux of WIMPs is estimated to be roughly 100 000 cm^{-2} s^{-1}. That is to say, 100 000 WIMPs flow through every square centimeter of the Earth's surface in one second. These particles continually rain down upon the Earth and, of course, mostly pass through without interaction – there are as many particles coming upward through the Earth as there are falling down from the sky. Mostly. Occasionally there is an encounter between a WIMP and an atomic nucleus, resulting in the exchange of several kilovolts of kinetic energy.

Direct detection involves the rare collisions of dark matter particles with target nuclei in terrestrial laboratories. At present, there are several dedicated dark matter searches with the goal only of detecting dark matter, unlike indirect detection observations with instruments such as IceCube or Fermi that have multiple scientific goals. The experiment is generally in an underground environment, such as an old mine or a deep tunnel, to reduce the background due to cosmic rays. Until recently, the most sensitive detectors were built around a target of semiconducting material, such as silicon or germanium, cooled to a fraction of a degree above absolute zero. When a dark matter particle strikes an atomic nucleus, the nucleus will recoil and generate sound waves (quantized sound waves called "phonons"). The sound waves heat the material and change its electrical conductivity; this change can be measured. Because of the necessary cooling these are called "cryogenic" dark matter detectors. So far, the most sensitive experiment using this method is the Cryogenic Dark Matter Search (CDMS), located in an old iron ore mine in Minnesota.

The principal development over the past five years has been the introduction of a new technology into the art of dark matter detection – detectors involving the detection of light pulses, luminescence, that occur when charged particles interact in the target material. Such detectors require arrays of photomultiplier tubes, as in the IceCube neutrino telescope. There are now two such large experiments in operation, both utilizing the noble gas, xenon, as the target: XENON100 in the Gran Sasso tunnel in Italy and LUX ("Large Underground Xenon" experiment) in the Homestake mine in South Dakota. They are designed to detect xenon nuclei that are occasionally scattered in a collision with a dark matter particle.

Xenon in liquid form is an excellent scintillating material. When a xenon nucleus is struck by an incoming particle, a neutron or a WIMP, it will recoil with an energy of a few keV and produce excitation and ionization in the surrounding medium. This energy ultimately appears as the emission of (ultraviolet) light at a characteristic wavelength of 178 nanometers; the xenon liquid is itself transparent to that radiation. Again the trick is to distinguish between a rare encounter between a xenon nucleus kicked by a dark matter particle and the much more frequent sources of ionizing radiation such as cosmic rays or impurities in the xenon itself. The ionizing radiation produces electrons that also cause scintillation, so it is important to discriminate between those interactions that produce electrons (the background) and those that involve nuclear recoil (the signal). In xenon detectors, this can be done by passing an electric current through the liquid xenon, which collects the electrons due to ionizing radiation in the top of the detector containing gaseous xenon. There the electrons are further accelerated by a strong electric field, where they produce scintillations due primarily to the ionizing component. The ratio of this electron signal to the primary scintillation signal in the liquid xenon can discriminate between the nuclear recoils and the electron recoils.

The first results of the XENON100 detector with 62 kilograms of xenon operating in Gran Sasso for 225 days were reported in 2012.[21] The first results of the LUX experiment (118 kg of xenon) operating for 85 days were reported in 2014.[22] The results for both detecters are consistent with the expected background,

7.4 The Science of Dark Matter Detection

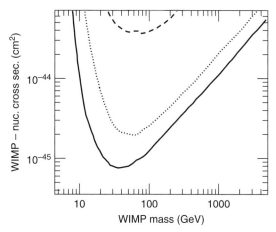

Figure 7.3. An exclusion diagram for a WIMP–nucleon scattering cross section and the mass deduced from upper limits in direct dark matter detection experiments. The lower curve (solid line) is that of the LUX experiment; the region above the curve is the range of cross section vs. mass excluded by this experiment. That is to say, a cross section just above 7×10^{-46} cm^2 for a WIMP mass of 60 GeV would be excluded by this experiment. The next-lowest curve (dotted line) is that of the XENON100 experiment, which clearly is less exclusive. The highest curve applies to the CDMS experiment. Clearly claims of detection in this experiment are ruled out by XENON and LUX, as is the DAMA result for spin-independent WIMP–nucleon scattering.[22]

i.e., no detection of WIMPs. The limit on the WIMP–nucleon spin-independent cross section for XENON was 2×10^{-45} cm^2 and that of LUX was about three times lower, as is shown by the exclusion diagram in Figure 7.3. The excluded region of the cross section vs. WIMP mass plane rules out other reported detections or hints of detections for lower-mass WIMPs, e.g., the DAMA and CoGent results for annually modulated WIMP signals (see Figure 7.3). The XENON detector is being upgraded to a target mass of one ton and LUX is continuing the search for 300 days, leading in both cases to substantially improved sensitivity. As of this date (September 2015) no WIMPs have been detected directly in terrestrial laboratories. Dark matter remains hypothetical.

7.4.3 The LHC and Dark Matter

The Large Hadron Collider (LHC), built and operated by CERN (the European Organization for Nuclear Research), is the most powerful particle accelerator in the world; in fact, it is the largest scientific instrument ever constructed. It is an underground circular tunnel 27 km in diameter beneath the Swiss–French border near Geneva. Its function is to accelerate hadrons (particles consisting of quarks and gluons, but in this case primarily protons) up to energies of 7 TeV (tera-electron volts, or 10^{12} eV) in two beams circling in opposite directions. Like two high-speed trains, the beams then smash into one another at a combined energy of 14 TeV. The tunnel is lined with more than 1600 electromagnets which keep the trains on their tracks. These electromagnets are cooled to two degrees above absolute zero so they are superconducting for maximum efficiency. Altogether it is a remarkable engineering achievement.

Unlike train collisions, the wreckage of particle collisions reveals much about the nature of the sub-atomic world. The LHC, working at half power, has already led to one fundamental discovery: the Higgs boson, which was the missing link in the standard model of particle physics; it is the Higgs field that gives mass to all particles. Now it is hoped that the LHC will provide experimental evidence for physics beyond the standard model – supersymmetry, for example. This is possible by detection of the hypothetical super partners among the debris of the collisions – those particles that differ by half-integral spin from the well-known fermions and bosons. And, of course, the lowest-mass superparticle is the leading candidate for WIMP dark matter. Can the LHC produce and detect dark matter?

To create particles in the collision, the rest-mass energy of the particles must be less than the colliding protons – 14 TeV in this case. This is consistent with the range of WIMP masses considered in direct detection experiments ($\approx$ 100 GeV). But then, how are the new and weakly interacting particles actually detected? The process depends upon a well-known physical law – the conservation of linear momentum. When two beams of protons collide, a spray of collision products is produced, mostly along the axis of the beams. But there are also particles produced that move transverse to the beam direction. That is permitted as long as the total transverse momentum (the momentum perpendicular to the beam axes) is zero; it must be, since there is no transverse momentum before the particles collide. So particles that interact very weakly can be detected by missing transverse momentum – by those collisions in which the total transverse momentum does not add up to zero.

At these high energies, such collisions are often distinguished by a *mono-jet* (Figure 7.4) – a packet of quarks and gluons all moving together. There is nothing seen coming out in the opposite direction, so there is missing momentum. However, this does not mean that a super particle has been discovered; it may be that other weakly interacting particles are carrying away the missing momentum – neutrinos, for example. In order to detect new particles, many such collisions must be seen and compared to the predictions of the standard model. An excess of such collisions over that predicted would imply new particles.

It might be the case that superparticles are found, confirming supersymmetry, but that they are not the dark matter. How can we decide if new particles have the properties required (the masses, the abundances, the interaction cross sections)

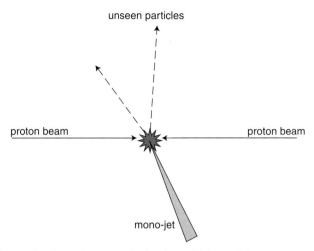

Figure 7.4. Production of a mono-jet in the collision of the two proton beams at the LHC. The particles necessary to conserve transverse momentum are not seen; they are weakly interacting and could be superparticles.

to actually comprise cosmic dark matter? It is only by verification from other experiments (the scintillation searches, for example) that any newly discovered particles can be confirmed as the dark matter. The LHC and other high-energy experiments can certainly narrow the range of the search, but in themselves such probes cannot definitively confirm new particles as cosmic dark matter. This requires corroboration by the ongoing dark matter search experiments – either direct or indirect.

7.5 The Sociology of Dark Matter Detection

The experimental detection of dark matter, direct or indirect, has become a fiercely competitive endeavor. This is understandable: to identify the primary material constituent of the world would be one of the greatest discoveries in the history of science – for any scientist it would be a memorable achievement to be involved in such a discovery. Considering the investment of money and careers in dark matter detection projects, the effort has also become a major industry with all of the vested interests of a major industry, and certainly the most pronounced goal of a vested interest is to make sure that the project goes on. Given the fact that non-detection does not constitute falsification, dark matter searches could, in principle, continue forever, achieving tighter and tighter limits on cross sections and particle masses. However, practically, there are limits set by the constraints of funding and the finite attention span of the relevant communities in pursuing projects that may lead to no positive results.

For now, the search continues and is likely to do so until the elusive particles are detected or until interest fades. The presence of many collaborators in several large competing projects has created an interesting social sphere: false hopes and rumors run rife and are pushed forward by the modern phenomena of internet blogs and rapid press releases. It seems that every few days in the popular scientific press we read of a new hint of dark matter, or of being on the "verge" or "threshold" of a discovery. It is a wide threshold. A Google search reveals that we have lingered on this threshold for twenty years. It is also the case that these "hints" are rarely retracted when it is subsequently found that they are exaggerated. It must be rather confusing for the interested public, many of whom probably think that the dark matter has been identified – at least until they see the next press release announcing that we are still hovering on the threshold. It would not be surprising if this were to lead to a certain amount of cynicism about progress in science.

With respect to indirect dark matter searches, it appears that any new unexpected signal is immediately attributed to dark matter annihilation. An example is the stir surrounding the first release of the AMS data on the positron fraction in cosmic rays: there was a flurry of excitement and press conferences suggesting that the dark matter had been found – all before a careful consideration of the likely astrophysical backgrounds. However, this demonstrates the advantage of the social process of science, in that with so many projects and groups involved there is always a host of devil's advocates sharpening their knives to cut apart any claimed detection.

The same is also true of direct dark matter searches. It is positive that there are so many different dark matter searches using several different methods. The competition helps to keep everyone honest; it pushes all groups forward and at the same time makes everyone cautious about exaggerated claims. Or at least it should. There have been several cases where close competition appears to have led to premature claims of success. An infamous such episode was that of the 2009 WIMP "detection" by the CDMS. For weeks ahead there were rumors that a major discovery would be announced by the group in mid December: the blogs were hyper-active generating excitement about the upcoming media event, to be revealed by several simultaneous seminars in various locations. Then when announcements came, it was a remarkable fizzle. The group had detected three events that were consistent with WIMP–nucleon scattering but also consistent with background. The events might have been WIMPs, but they might not have been – hardly a resounding discovery.

This furious activity was not long before the new generation of scintillation dark matter detectors – the xenon-based technology – began to come online. It appeared to some critics that the group involved in cryogenic detection was attempting a last-ditch effort to score a success before these more sensitive devices began

7.5 The Sociology of Dark Matter Detection

producing results. When the negative XENON and LUX results were published, it was immediately evident that the upper limits on the WIMP–nucleon cross section were far below those implied by the suggested detection – that is, if the CDMS had really detected WIMPs, hundreds of such events would have been seen by XENON and LUX in one year. CDMS had detected no WIMPs.[23]

The scientists involved in the CDMS collaboration are certainly honest and, no doubt, believed that they had a probable detection. But this is an example of being overly optimistic – of wanting a positive result too badly. In such a quest, it is better to err on the side of caution – in that way a group becomes more credible when it does make a claim of detection.

This is also true of the famous DAMA/LIBRA WIMP detection. DAMA (also located in the Gran Sasso tunnel) uses solid scintillation detectors consisting of sodium iodide. The experiment gets around the problem of background by looking for an annual modulation of the entire signal – possible real WIMP–nucleon scatterings as well as background. A modulation in the WIMP signal can, in principle, be caused by the annual variation of the mean WIMP wind blowing on the Earth due to its motion about the Sun. Such a wind would blow stronger in June, when the velocity vector of the Earth about the Sun was more closely aligned with that of the solar motion about the Galactic Center, than in December, when the opposite is true. The group has presented convincing evidence that a modulation with this signature is indeed present in the total signal. However, it is not obvious that this modulation is due to the WIMP wind.[24] Again, the estimate of the spin-independent cross section is significantly above the upper limits implied by the more sensitive XENON and LUX experiments. Yet the DAMA group aggressively promotes their observed modulation as a WIMP detection; it would be more impressive if they expressed some doubt rather than absolute certainty about their result, but a degree of self-promotion is apparently in the nature of the WIMP detection industry.

To attempt, in a deep mine, to detect several events per year in a hundred kilograms of silicon or xenon is much more difficult than finding the proverbial needle in the haystack – yet this exercise is now performed routinely. Even in the absence of actual detections it is a remarkable achievement and, no doubt, positive technological spin-offs will follow. But this, of course, is not the motivation of those involved: the goal that drives the scientist is to identify the material content of the world and to move particle physics beyond its standard model – to do high-energy physics with low-energy experiments. In the next chapter I will argue that this is not likely to be successful.

8
MOND

MOND (modified Newtonian dynamics) posits a drastically different approach to the problem of the observed mass discrepancies in astronomical objects. The idea is based upon a simple proposition: if the only evidence for dark matter on astronomical scales is its putative global dynamical or gravitational effects, then its presumed existence is not independent of the laws of dynamics or gravitation on those scales – that so long as no candidate dark matter objects or particles have been identified, then it is legitimate to look for alternative solutions to the discrepancy in modifications of Newtonian dynamics or gravity. Such a point of view is hardly radical at all but would seem to be a reasonable scientific approach. And yet, the very mention of "MOND" evokes strong reactions among astrophysicists and cosmologists; most of that reaction is not benign.

MOND is an acceleration-based modification of Newtonian dynamics or gravity. Now, many years later, it has been realized that in the deep MOND limit, for accelerations below a critical value, the theory reflects a very basic symmetry – a symmetry that is already evident in galaxy phenomenology – and that is where I begin.

8.1 Galaxy Phenomenology Reveals a Symmetry Principle

There are three aspects of galaxy phenomenology that seem unnatural in the context of cold dark matter. The first is that the rotation velocity beyond the visible galaxy approaches a constant fixed value; rotation curves are asymptotically flat. This flatness of rotation curves, shown so vividly by the example plotted in Figure 7.2, is a general feature of spiral galaxies – at least in those cases where there are no complications, such as nearby interacting companion galaxies or large distortions (warps) of the gas layer.

In galaxies of high surface brightness, there may be a gradual decline in the visible disk, but then, with increasing distance, the rotational velocity approaches

8.1 Galaxy Phenomenology Reveals a Symmetry Principle

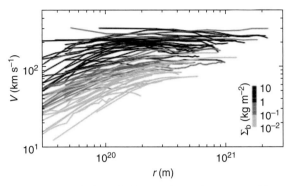

Figure 8.1. This is a compilation of observed rotation curves illustrating the fact that the rotation curves become constant beyond the visible disk (asymptotic flatness). In this compilation the surface density generally increases vertically; upper curves are for galaxies with the highest surface brightness. We see that for the low-surface-brightness objects the rotation velocity typically increases to the constant asymptotically constant value, whereas for the higher surface brightness objects the rotation velocity is constant throughout or decreases to the constant value. This plot is given in the review by Benoit Famaey and Stacy McGaugh.[1]

its constant value. For low-surface-brightness galaxies, there is often a slow rise throughout the visible disk, but at larger distances the rotation velocity rises to a fixed value. This behavior is shown by the compilation of galaxy rotation curves in Figure 8.1; it is a general phenomenon.[1]

Rotation curves are asymptotically flat as far out as they have been measured. They do not slowly rise; they do not slowly decline. This simple fact has never been fully appreciated by the majority of astronomers and cosmologists. In terms of dark halos it means that a very specific density distribution must apply in all objects: specifically, the density of dark matter must decline as $1/r^2$ to provide the necessary Newtonian gravitational field; there can be no variation in this density law from object to object in spite of the different dynamical histories of different galaxies.[2]

The second remarkable aspect is that the value of that asymptotic constant rotational velocity, far from the central galaxy, depends precisely upon the directly observable baryonic mass of that galaxy. In the context of dark matter, this would mean that there is an exact correlation between the baryonic mass of the relatively small object in the center and the rotation velocity of particles in circular orbits in the vast extended dark halo. The form of that correlation is $M \propto V^4$ (see Figure 8.5), and the scatter about the average relationship is consistent with observational error: there is no intrinsic scatter in the relationship itself. The correlation holds over a range of galaxy types and baryonic masses – from low-mass, gas-dominated dwarf galaxies up to massive large spirals with substantial central spheroidal bulges. In spite of the very different histories of formation, merging, interaction and evolution, galaxies lie on this exact correlation between the observable mass of the visible object and the circular velocity in the great extended dark halo. How has nature contrived to reconcile this essential aspect of the gravitational field of an extensive dark halo to the mass of the central baryonic object – and with such precision?

The third fact, discordant with CDM, is that the discrepancy between the detectable baryonic mass and the traditional Newtonian mass appears below a critical acceleration – i.e., not at large distance per se but at low accelerations. The appearance of a critical acceleration became clearly evident only when the observations of galaxy rotation curves achieved the precision sufficient to map the discrepancy as a function of distance or centripetal acceleration. But it is now undeniably clear that there exists a definite value of the acceleration – on the order of 10^{-10} m/s^2 – below which the discrepancy in galaxies appears (see Figure 8.2). Again, with respect to the CDM paradigm, this is puzzling because there is no such acceleration scale that is intrinsic to dark halos, at least not the kind of halos that emerge from cosmic computer simulations with millions of point masses representing cold dark matter.

These three properties of galaxy phenomenology imply that the unifying aspect may lie in basic physical law and not in the random histories of galaxies. Moreover, given that the basis of modern physics rests in symmetry principles, the phenomenology suggests a specific kind of symmetry.

In geometry, we all have an idea of what symmetry means. A sphere, for example, has a high degree of symmetry: from any point of view, it appears the same.[3] A more precise definition of spherical symmetry is that the form of the sphere does not change after rotation about any axis. This an example of a general operational definition of symmetry going beyond geometry: a symmetry leaves some property of a system intact after performing a transformation – a mathematical operation. Important operations in physics are translation (moving a system from one place or time to another) and rotation about some axis. The classical equations of motion, for example Newton's laws, do not change with (i.e., are invariant to) a translation in space and time. In the early part of the twentieth century, the great mathematician Emmy Noether[4] proved that such symmetries can imply the presence of conserved quantities; for example, invariance to translation in space–time implies conservation of linear momentum and energy, and invariance to rotation implies conservation of angular momentum. One of the important symmetries in physics is the so-called Lorentz symmetry[5] – that corresponding to rotations and changes of velocity. The profound consequence of this symmetry is that the speed of light is the same in frames of reference moving with constant velocity relative to one another; the velocity of the light is invariant to the motion of the frame in which it is measured. This forms the basis for special

8.1 Galaxy Phenomenology Reveals a Symmetry Principle

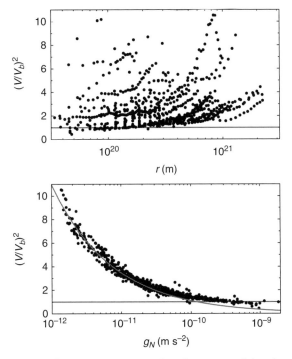

Figure 8.2. The mass discrepancy measured as the square of the observed rotation velocity in terms of that expected from baryons alone (acceleration). The vertical axis for each panel is the mass discrepancy as measured by the square of the observed rotation velocity in terms of that expected from the baryons alone. The individual points are taken from observations of a number of rotation curves at independent positions. The discrepancies are plotted in the upper panel as a function of radial distance within the galaxies, and in the lower panel as a function of the Newtonian acceleration (g_N) in units of m/s². The result clearly shows the absence of any dependence on distance (the discrepancy is not larger at larger distances) and the correlation of the discrepancy with acceleration. In the lower panel the horizontal line shows the expectation from Newtonian dynamics, and the curve, more nearly following the points, shows the expectation from the deep MOND limit. We notice that the two curves cross near an acceleration of 10^{-10} m/s²; this marks the transition between Newtonian and modified dynamics. The figure is reproduced with the permission of Stacy McGaugh.

relativity and implies that there are no preferred inertial frames where physical laws take their most simple form.

What symmetry, or invariance principle, might be suggested by observed galaxy phenomenology? The essential clue is provided by the constant rotation velocity in the low-acceleration regions of spiral galaxies. Velocity has units of space divided by time, so if we multiply the spatial coordinate r and the time coordinate t by the same factor b (this would be the simple operation $r' = br$, $t' = bt$) then the velocity remains the same (see Figure 8.3 for a simple illustration of this principle).

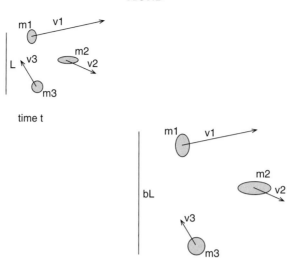

Figure 8.3. Space–time scale invariance for a small galaxy group. Here space and time are expanded by a factor of b, but the velocity of the individual galaxies remains the same. The solution of the equations of motion has the same form for both systems related by this transformation; that is, the form of $r'(t')$ is the same as that of $r(t)$. This plot is reproduced with the permission of Moti Milgrom.

This suggests that the symmetry principle leading to flat rotation curves is that of *space–time scale invariance*. The dynamical equations should remain the same after expanding (or contracting) space and time by the same factor. We note that this symmetry does not correspond to a global change of units (say, centimeters to inches or seconds to hours); of course, all physics should be independent of this sort of trivial transformation. It is an actual stretching or contraction of intervals of space and time in which the dimensioned constants of a system G and specific masses, and any other dimensioned constants that might be introduced in a modification of Newtonian gravity, remain the same. Milgrom[6] has shown that applying such a symmetry principle to a purely gravitational system requires that the rotation velocity in the system V does not depend upon the distance r, and that there exists a definite relationship between the central mass and the value of that constant rotation velocity ($M \propto V^\alpha$). The equation relating acceleration to attraction in usual Newtonian gravity ($1/r^2$ attraction) does not obey such a principle – it is not space–time scale invariant, so this symmetry principle describes a modification of Newtonian gravity.[7]

This symmetry alone does not specify the type of modification. Any modification leading to a constant circular velocity can result from such an invariance principle. For example, it is possible to specify that $1/r^2$ breaks down and becomes $1/r$ attraction beyond some scale r_0. Then the relation for gravitational acceleration becomes invariant when space and time are expanded or contracted by a factor of b.[8] In this case the equation of motion would imply a mass-rotation velocity relation about a point mass of the form of $M \propto V^2$.

This is not observed. If, however, we specify that the modification occurs below a *critical acceleration* a_0, the equation of motion becomes $g = \sqrt{Ga_0M}/r$, which is space–time scale invariant and leads directly to the observed mass–velocity relation of the form $M \propto v^4$ – to modified Newtonian dynamics or MOND. This approach, as discussed by Milgrom, is an elegant route to MOND. Whether or not the symmetry is fundamental (in the sense of Noether) or incidental is another question, but it is certainly useful (for example, the scaling operation applied to any deep MOND solution gives another solution). However, the reason why the idea is taken seriously by a number of astronomers and physicists is that MOND, methodologically, is a highly successful algorithm that embodies a large range of galaxy phenomenology.

8.2 An Empirically Based Algorithm

Milgrom's unique insight was to realize, early on, before it was confirmed observationally, that only a modification attached to an acceleration scale (not a length scale) could explain the existent observations: the presence of small galaxies with large discrepancies between the detectable and dynamical mass and the form of the observed correlation between the detectable baryonic mass of galaxies and their asymptotic constant rotation velocity – the Tully–Fisher relation, $M \propto V^4$.

Viewed simply, MOND is an algorithm that, with a single additional physical constant having units of acceleration, allows one to calculate the effective gravitational force in an astronomical object from the observed distribution of baryonic matter. And it works remarkably well. This is evidenced primarily by use of the algorithm to calculate the rotation curves of disk galaxies from the observed distribution of baryons – stars and gas; the agreement with observed rotation curves is in most cases quite precise, even with respect to details. The existence of such a successful algorithm is problematic for cold dark matter because this is not something that dissipationless dark matter can naturally do; it would seem to require a coupling between dark and baryonic matter that is at odds with the perceived properties of CDM.

8.2.1 Galaxy Rotation Curves

The algorithm can be written as a modification of Newtonian gravitational attraction. Here the relation between the true gravitational attraction g and the

Newtonian gravitational attraction g_N is given by $g = \sqrt{a_0 g_N}$ whenever $g_N \ll a_0$. In the limit of large accelerations $g_N \gg a_0$, we have $g = g_N$ as in standard dynamics. A more general relationship between the actual acceleration and the Newtonian acceleration would be $g = g_N \nu(g_N/a_0)$, where the function $\nu(x)$ interpolates between the two limits; i.e., $\nu(x) = 1$ when $x \gg 1$ and $\nu(x) = x$ when $x \ll 1$. The actual form of ν is not very critical when considering galaxy phenomenology.

Given the observed distribution of starlight and gas in disk galaxies, we can calculate the Newtonian force law g_N within that galaxy (assuming of course a mass-to-light ratio for the stellar component). Then we can apply the simple MOND formula to calculate the "true" gravitational acceleration g and hence the rotation curve of the galaxy. The results are shown in Figure 8.4 for a sample of 15 spiral galaxies; the points are the observations, the dotted and dashed curves are the Newtonian rotation curves of the baryonic components stars and gas, the long-dashed curves show the Newtonian rotation curves of a central bulge component when present, and the solid curves are the modified rotation curves. The mass-to-light ratio is the only adjustable parameter (different for each galaxy), and of course there is the universal acceleration constant a_0 that here is set to be 1.2×10^{-10} m/s^2. We see that the match between the observed and MOND rotation curves is very good in most cases (it is difficult to discern the solid curve from the points for several objects). Moreover, the required mass-to-light ratios are reasonable – values of one to three in solar units.[9] These rotation curves are a sub-sample of roughly 100 objects where the agreement, for about 90 objects, is clear-cut.

A further example, shown in Figure 8.5, illustrates a very significant aspect of galaxy rotation curves that MOND predicts, but that dark matter can not even explain. UGC 7524 is a dwarf, low surface brightness galaxy[10] with a large discrepancy between the detectable and Newtonian dynamical mass. The top panel of this figure is the logarithm of the surface density in stars and gas plotted against the radius in thousands of parsecs (the stellar surface density is determined from the surface brightness by assuming a mass-to-light ratio of 1.6 in solar units). The bottom panel shows the rotation curve as measured in the 21-cm line of neutral hydrogen (points with error bars) as well as the Newtonian curves of the luminous stellar component (dotted curve) and the gaseous disk (dashed curve). We can see that for both stars and gas there is a significant enhancement of the surface density between 1.5 and 2.0 kiloparsecs and that there is a corresponding feature in the Newtonian rotation curves. However, there is also a feature at this same radius in the total observed rotation velocity, even though there is a significant discrepancy between the Newtonian and detectable mass. The observed rotation curve perfectly reflects the details in the observed baryonic mass distribution even

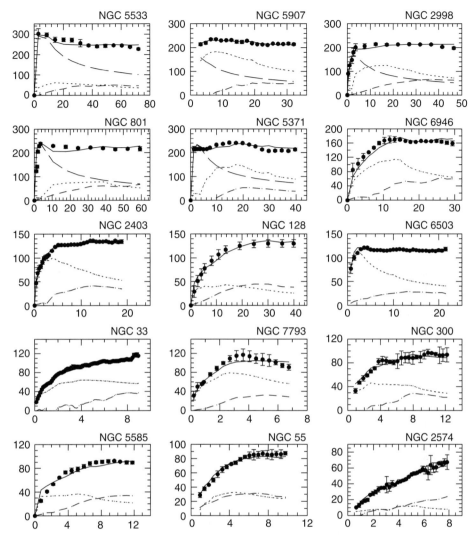

Figure 8.4. A sample of observed rotation curves (points with error bars) along with the Newtonian curves of the stellar disk (dotted), the gaseous disks (short-dashed) and in some cases, a central bulge (long-dashed). The solid curve is the MOND rotation curve determined from the Newtonian curves using the MOND algorithm. Notice that the third galaxy down on the right is NGC 6503, the object with the very extended flat rotation curve shown in Figure 7.2. See R. Sanders.[9]

though the object is, supposedly, dominated by dark matter. The solid curve is the curve calculated using the MOND algorithm that matches the observations.

This illustrates an empirical point that has been emphasized previously:[11] *For every feature in the surface brightness distribution (or gas surface-density*

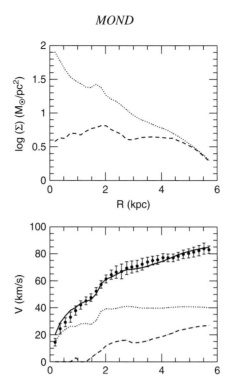

Figure 8.5. Top: the mass surface density of stars and gas (dotted and dashed curves) as a function of radius for the low surface brightness galaxy UGC 7524. Bottom: the corresponding Newtonian and MOND rotation curves (dotted, dashed, solid) and the observed rotation curve (points).

distribution) *there is a corresponding feature in the observed rotation curve and vice versa*. With dark matter this seems quite unnatural. How is it that the distribution of dark matter, a weakly interacting component with a high velocity dispersion, could so perfectly mimic the baryonic matter distribution? How can dark matter maintain sharp features in a rotation curve when the particles do not move on circular orbits but mix over different radii? With MOND it is the expected result. What you see is all there is.

8.2.2 The Baryonic Tully–Fisher Relationship

With MOND, equating the centripetal acceleration of a particle in a circular orbit (i.e., $g = V^2/r$) to the acceleration of gravity due to a point mass M (a reasonable assumption well beyond the visible disk), then in the low-acceleration regime, we find that $V^4 = Ga_0M$. That is to say, the asymptotic circular velocity is constant as observed in spiral galaxies and there is a relation between the true (baryonic)

8.2 An Empirically Based Algorithm 105

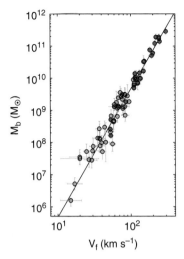

Figure 8.6. The baryonic Tully–Fisher relation. This is the baryonic mass plotted against the rotation velocity for a sample of spiral galaxies. The solid points are galaxies in which the stellar mass dominates (solid points). The stellar mass is determined from the observed 3.6-micron flux combined with population synthesis models to estimate a mass for the stellar disk. The grey points are galaxies (generally dwarfs) in which the gas mass dominates; the baryonic mass in gas is measured directly from 21-cm line observations of neutral hydrogen. The line is not a fit but is the expectation from MOND, with a_0 determined from rotation curve fitting. This plot is reproduced with the permission of Stacy McGaugh.

mass and the circular velocity of the form $M \propto V^4$. This simple relationship forms the basis of the baryonic Tully–Fisher relation.

The observations are shown in Figure 8.6, where the baryonic mass of a sample of spiral galaxies is plotted against the asymptotic circular velocity – the flat part of the rotation curve beyond the visible component. The mass includes the stellar component (where the luminosity has been converted into stellar mass using reasonable stellar population models[12]) as well as the contribution of the gaseous component (significant in low-luminosity galaxies). The solid line is not a fit but is the robust prediction of MOND with a dispersion of 20% about a mean value of $a_0 = 1.2 \times 10^{-10}$ m/s^2, as implied by fits to galaxy rotation curves. Thus MOND not only sets the slope of the relationship (4 for the power law) but also the normalization via a_0 – the same critical acceleration as that required to fit galaxy rotation curves. This point has been repeatedly emphasized by Stacy McGaugh: The scatter observed in this relationship is entirely consistent with observational errors, which is to say that no intrinsic scatter is detectable.

In cosmological simulations with CDM, halos of galactic mass form at about the same cosmic time and have roughly the same average density (with considerable scatter). Combined with the Newtonian virial theorem (a relation between the velocity dispersion, mass and size of a self-gravitating object in equilibrium), the form of the velocity–mass relation for halos is $M \propto v^3$ (putative dark halos are not supported by rotation but by the random motion of the halo particles; this is comparable to the circular velocity that would enter into the Tully–Fisher relation). To bring this expectation into agreement with the observations, it is supposed that galaxies lose a fraction of their baryons, which is greater for low-mass galaxies (recall that most baryons in galaxies appear to be "missing"). This must happen in such a way as to greatly reduce, not increase, the scatter in the halo mass–velocity relationship.

Recently, observations of the velocity-mass relation have extended down to very low accelerations – a few percent of a_0 – and to large distances from the parent galaxy. This observation makes use of weak gravitational lensing, in which the lens (a large foreground galaxy) systematically distorts the images of faint background galaxies – galaxy–galaxy lensing. Since the effect is very small, the technique is statistical: the background distortions are measured over a number of lenses and the lenses can be binned by luminosity.[13] The "dark halos" are found to extend to large distance (hundreds of thousands of light years) and are modeled as isothermal spheres – objects characterized by a constant velocity dispersion (a spread in random velocity); this velocity dispersion is proportional to the constant asymptotic rotation velocity in the "halo."

Such observations provide the opportunity to test MOND (and dark halos) down to very low accelerations and to large extent. The results are shown in Figure 8.7 as a plot of the velocity dispersion against the luminosity of galaxies averaged in bins. The lines are the predictions of MOND for both red and blue galaxies, where the mass-to-light ratios range from one to six in solar units.[14] Keeping in mind that MOND predicts both the slope and the scaling of the relationship, we note that this prediction is validated over orders of magnitude in acceleration. The mean Newtonian mass to observable mass ratio extends to 40 or 50 in solar units. This means that the baryonic mass in these systems is a few percent of the putative dark mass; the extent of the visible mass is also a few percent of that of the "dark halo." And yet, over these very large scales, the rotation velocity that supposedly is set by the halo is perfectly correlated with the miniscule baryonic mass at the center of this supposedly large configuration.

There has been much work in attempting to accomplish a similar baryonic mass–rotation velocity relationship by so-called "semi-analytic galaxy formation" programs: an exercise characterized by parameterized modeling of "complicated baryonic physics" or "gastrophysics" to push the expectations to conform to observations. With fine-tuning of sufficiently numerous free parameters it is not surprising that it can provide a general correlation (although not the small scatter),

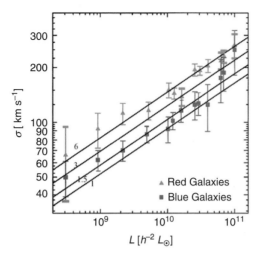

Figure 8.7. The Tully–Fisher relation extended to very low accelerations and very large distances by weak gravitational lensing observations of galaxy–galaxy lensing. The lines show the MOND predictions for the indicated mass-to-light ratios and are relevant to both red and blue galaxies (triangles and squares). Reproduced by permission of Moti Milgrom.

but this is in stark contrast with MOND, in which the precise relation follows not from the details of galaxy formation but from existent physics.

This predictive power on the scale of galaxies is notably absent with CDM; thus, the mere existence of such a successful algorithm constitutes a severe challenge for CDM or any supposed dissipationless dark matter that clusters on the scale of galaxies. In fact, one might argue that MOND provides a falsification of CDM as it is perceived to be, and therefore impacts on the entire cosmological paradigm.

8.2.3 A Critical Surface Density

It has been argued that confrontation of MOND with galaxy rotation curves and the baryonic Tully–Fisher relationship are not true predictions because MOND is "designed" to reproduce these features. That is to say, flat rotation curves and a Tully–Fisher relation of the observed form are part of the propositional basis of MOND. Of course, none of the detailed data on rotation curves existed when MOND was proposed, and there was also controversy about the true slope of the Tully–Fisher relationship, so one could answer that designing a theory to explain data not yet taken is called prediction. But there is one aspect of MOND that unquestionably falls into the category of true prediction: the presence of a large discrepancy in low-surface-brightness galaxies, and correspondingly, a small discrepancy in high-surface-brightness objects.

Since both the Newtonian gravitational acceleration and the surface density of matter in a system (the mass per unit surface) vary as mass divided by radius squared, it is evident that the MOND critical acceleration can also be written as a surface density: approximately a_0/G. This turns out to be about 300 solar masses per square light year (a few tenths of a gram per square centimeter). The significance of this value in the context of MOND is that systems with lower surface density should exhibit a larger discrepancy between the dynamical and the observable baryonic mass; systems with higher surface density are in the Newtonian regime and there should be no significant discrepancy within the high surface brightness regions. In so far as surface brightness (solar luminosities per square light year) is proportional to surface density, this means that there will be a large discrepancy in faint low-surface-brightness objects (in traditional language – much more dark matter in low-surface-brightness galaxies). This was an initial prediction of MOND which has been brilliantly verified.

It is well known that the Newtonian dynamical mass of compact bright objects such as globular star clusters is well accounted for by the luminous mass in stars – there is no discrepancy implied by the Newtonian dynamical mass-to-light ratios, at least not within the bright visible object.[15] On the other hand, very faint diffuse objects, such as the dwarf spheroidal galaxies in the neighborhood of the Milky Way, have a very large discrepancy, with Newtonian mass-to-light ratios approaching 100 in several cases – an observational result obtained *after* MOND was proposed. Viewed in terms of conventional dynamics, they are dominated by dark matter within the visible object. This dichotomy is immediately understandable in terms of an acceleration-based modification such as MOND.

Strong gravitational lensing, the formation of multiple images or an Einstein ring of a distant source by an individual foreground galaxy, probes the high-acceleration regime of these lenses (see Figure 8.5). This is because of the fact that there is a critical surface density in the lens for strong lensing given by $(cH_0/G)F(z_l, z_s)$. This happens to be the MOND critical surface density multiplied by a dimensionless function F of the source redshift z_s and the lens redshift z_l; the numerical value of this function is typically 5–10 for cosmic lenses. In other words, strong lensing can only occur in regions where the acceleration is greater than a_0. This means that there should be no significant discrepancy (little dark matter) revealed by isolated strong gravitational lenses.

A reasonably large sample of strong gravitational lenses is provided by the SLACS (Sloan Lens ACS Survey) sample.[17] The lenses are mostly relatively gas-free spheroidal galaxies (not star-forming disk galaxies) and the form of the lensed images is often complete or near-complete Einstein rings, indicating near alignment of the background source with the lens (as in Figure 8.8). A plot of the mass projected within the Einstein ring (using standard general relativity)

8.2 An Empirically Based Algorithm

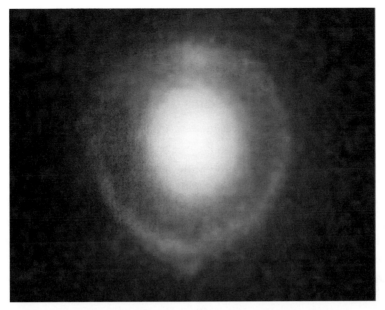

Figure 8.8. A strong gravitational lens. The lens itself is a massive foreground spheroidal galaxy. The image of the source, a more distant blue galaxy that is closely aligned with the lensing galaxy, forms an almost perfect ring – an Einstein ring – due to the gravitational field of the lens. The angular size of the ring is related to the average surface density of matter within the ring. This is from the SLACS lens survey.

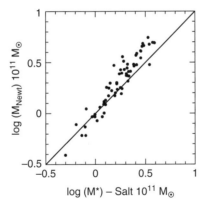

Figure 8.9. The logarithm of the GR lensing mass of SLACS lenses plotted against that of the stellar mass projected within the Einstein ring. The stellar mass of these objects (given by Auger et al.[16]) was estimated by assuming that newly formed stars have a commonly used power law distribution by mass – the Salpeter initial mass function. Both the stellar mass and the lensing mass are given in units of 10^{11} $M_\odot$ and the equality line is shown.

compared with the observed projected stellar mass[18] is shown in Figure 8.9. It is evident that there is no significant discrepancy within the inner high surface brightness regions of these lensing galaxies as predicted by MOND.

8.3 Cosmology and the Critical Acceleration

Can we identify a natural value of a_0 in the world? Does cosmology provide such a preferred acceleration? The acceleration of gravity for the entire enclosed Universe at the Hubble distance (c/H_0) is approximately cH_0. Moreover, we apparently live in a de Sitter universe, or at least in a universe that is becoming a pure de Sitter universe – that is to say, an accelerating universe, as is evidenced by the distant supernovae. The cosmological acceleration increases linearly with distance from any object (such as the Milky Way), that is $a \sim (\Lambda/3)r$, where Λ is the value of the cosmological constant ($\Lambda/3 \sim H_0^2$). As the distance approaches the de Sitter event horizon, $c/\sqrt{\Lambda}$), this acceleration approaches its maximum value, near $c\sqrt{\Lambda}$. These two considerations would appear to deliver a cosmic value on the order of a_0,[19] and are both on the order of 10^{-10} m/s².

Could it be that this cosmologically preferred acceleration enters into the dynamics of systems on a small scale as compared to the Hubble radius? If so, then MOND could be an effect of cosmological expansion on the dynamics of local systems. There is certainly one local phenomenon that would be modified at accelerations below this preferred cosmic value in a de Sitter universe; that is, the Unruh radiation.[20] A particle in uniform linear acceleration, a, finds itself immersed in a uniform radiation field with a black body spectrum; the temperature of this radiation is proportional to the acceleration, i.e., $T = Ka$, where $K = \hbar/(2\pi ck_B)$ and k_B is the Boltzmann constant. This is similar to a black hole where an observer far from the event horizon of the hole also sees black body radiation emitted from the hole, the Hawking radiation, with temperature given by $T = Kg_h$, where g_h is gravitational acceleration at the horizon.

A de Sitter universe also possesses an event horizon at the Hubble radius; an observer can have no contact or awareness of any events beyond the event horizon – space beyond this horizon is cut off from any interior observer. And just as for a black hole, any observer also sees black body radiation emitted by the de Sitter event horizon; the temperature of this radiation is here proportional to the cosmic acceleration: $T = Ka_0$.[21]

But what about a uniformly accelerating observer within a non-trivial universe – a de Sitter universe? In this case, the Unruh radiation will be modified by the presence of this cosmic background acceleration: the temperature becomes $T = K\sqrt{a^2 + a_0^2}$. Could this have any effect on the dynamics of the accelerating particle – perhaps upon the inertial reaction? The Unruh radiation is too miniscule to be directly implicated in any dynamical effects such as inertia, but suppose the temperature traces the inertia of an accelerating particle – that the inertia is proportional to this temperature in de Sitter space less the temperature seen by a non-accelerating observer, i.e., to $F/m \propto K[\sqrt{a^2 + a_0^2} - a_0]$. If this were true, then the relation between force and acceleration would be modified from its Newtonian form: Newton's second law would no longer be $F = ma$ but $F = ma^2/a_0$ in the low-acceleration limit. There would be a transition from force proportional to acceleration to force proportional to a^2 occurring at accelerations below a_0. And this is precisely the form of modified Newtonian dynamics[22] expressed as a modification of inertia. In the limit where a_0 approaches zero, all dynamics would appear in their original Newtonian form.

This is not a theory but an example of how a modification of Newtonian dynamics might be related to the cosmic acceleration – as a modification of particle inertia in a de Sitter world. This issue is unclear, but the empirical success of the MOND algorithm on the scale of galaxies is very clear.

8.4 Problems with MOND

The principal empirical problem with MOND concerns clusters of galaxies – the systems in which the dark matter problem was first identified. It has long been recognized that MOND does not remove the entire mass discrepancy in clusters of galaxies.[23] A major observable component of galaxy clusters is hot gas that emits X-rays. Using observations obtained by X-ray observatories on space platforms, it is possible to map the density and temperature distributions and mass of this hot gas out to millions of light years from the cluster center. Assuming the gas is in hydrostatic equilibrium, astronomers can then calculate the Newtonian dynamical mass and, applying the MOND version of the equation of hydrostatic equilibrium, also the dynamical mass in the context of modified dynamics. In the Newtonian calculation the dynamical mass is typically six or seven times larger than the observable mass in hot gas and in the stellar component of the galaxies – compatible with the universal dark-to-baryonic ratio in the concordance cosmology. With MOND that factor is reduced to a factor of two or three, which is to say that MOND reduces but does not eliminate the mass discrepancy in rich clusters. The problem for MOND is that in the central parts of clusters the acceleration is greater than a_0, so MOND predicts that there should be no discrepancy between the observed and dynamical mass in the cores of the clusters, contrary to observations.

Formally, this observation does not falsify MOND. It would do if MOND predicted less mass than was observed; we may always identify more mass but we cannot make observed mass go away. So in fact, one could look at this problem as a bold prediction: MOND predicts that more mass will be found in clusters. There have been several suggestions: ordinary neutrinos with masses of 2 eV,[24] sterile (completely non-interacting) neutrinos with a mass of 11 eV,[25] small compact clouds[26] and low-surface-brightness galaxies.[27] These last two possibilities are in the form of unseen (or difficult to see) baryonic matter, but there are more than enough unseen baryons to make up the difference.

One thing is certain: whatever is missing, it must be dissipationless (non-collisional), like particle dark matter. This is evident from the observations of the famous "bullet" cluster where, using a combination of X-ray observations and gravitational lensing, we see the collisionless components (the galaxies and the dark matter) passing through one another while the collisional gas remains in the middle where the clusters have collided. For now, this remains an unresolved issue for MOND or for any modified-gravity replacement of dark matter.

Most of the original criticisms of MOND concerned matters of principle, such as the fact that systems consisting of several interacting masses do not conserve their linear and angular momentum. These issues were addressed by the non-relativistic field theory of Jacob Bekenstein in collaboration with Milgrom – a theory of MOND as a modification of the Newtonian gravitational field.[28] This theory does embody these cherished conservation principles, but as a non-relativistic theory it cannot address phenomena such as gravitational lensing and the formation of structure in a cosmological context. The theory is clearly incomplete.

This pinpoints the principal problem for MOND: the absence of a generally accepted relativistic extension of the idea. Without a relativistic theory, it is not possible to address cosmology and the formation of structure – those large-scale issues in which the standard paradigm does so well. In the context of general relativistic cosmology, it is possible to derive the Friedmann equations using only Newtonian gravity and dynamics as it applies to an expanding spherical region of the Universe. We can try the same with MOND – applying the modified dynamics to a finite spherical region – but this leads to contradictions. Uniform expansion is not possible: the expansion history of any spherical region depends upon the physical size of the region, so one cannot define a dimensionless scale factor to describe the expansion of the Universe as a whole.[29] The MOND algorithm can be applied to fluctuations about the average density, but then we find that the growth of structure is catastrophically rapid – MOND is too effective in promoting gravitational collapse (of course, the growth of structure is also related to the question of the constancy of a_0 with cosmic time – an issue that clearly requires a more complete theory in a cosmological context).

In the last decade, several relativistic theories of MOND have been proposed, and these have the potential to address cosmological problems. Most of these

theories require additional fields (degrees of freedom) such as one or more
scalar fields and/or vector fields. These relativistic theories have two aspects in
common: they reduce to the simple MOND algorithm in the non-relativistic limit
and they produce gravitational lensing equivalent to that of general relativity
with dark matter. The first consistent theory was that of Jacob Bekenstein –
tensor–vector–scalar gravity or TeVeS – a theory that involves two fields (or
degrees of freedom) in addition to the gravitational field of general relativity (the
space–time metric).[30] There are now theories that add more of these fields so they
are more complicated (and less aesthetic) than general relativity. In fact, there are
too many theories; only one, at most, is likely to be correct and share all of the
successes of general relativity on a smaller scale. It is unclear how these theories
address cosmological phenomena such as the pattern of anisotropies in the CMB.

On the grounds of aesthetics, it might be argued that general relativity is to be
preferred to more complicated theories involving additional degrees of freedom or
fields. General relativity, as a theory, has the advantage of being more economical
and elegant. But the ΛCDM paradigm requires more than general relativity. It
requires at least two additional sources – dark energy and dark matter – for
which there is no evidence independent of astronomical observations, with energy
densities that are set by these observations and that happen to coincide at the
present epoch.

In any case, the existence of an algorithm that so precisely predicts the force
in galaxies from the observed distribution of baryonic matter is devastating
for dark matter that clusters on the scale of galaxies – CDM. The implied
relationship between baryons and dark matter seems very strange: baryons behave
very differently from dark matter; they dissipate and collapse to the center of a
self-gravitating system; they are blown out by supernovae; they are left behind in
collisions between galaxies or clusters. In view of all these random processes, it
is difficult to comprehend how baryons could be such a perfect tracer of the dark
matter density distribution or how the circular velocity in the very extended halo
could be so well correlated with the baryonic mass (no intrinsic scatter). To believe
that poorly understood mechanisms such as "feedback" or "self-regulation" could
recover such tight correlations demonstrates a remarkably naive acceptance of the
cosmological paradigm.

And finally there is the ubiquitous appearance of $a_0 \approx cH_0 \approx c\sqrt{\Lambda}$. How is it
that dark halos, which embody no intrinsic acceleration scale, account for the fact
that a_0 is the acceleration below which the discrepancies appear in galaxies, that
a_0 determines the normalization of the Tully–Fisher relation in spiral galaxies as
well as the corresponding Faber–Jackson relation for elliptical galaxies, and that
a_0 defines a critical surface brightness below which the discrepancy is present and
above which it is absent?

This body of evidence, which is mostly not considered as an argument against CDM, in fact constitutes a profound case against CDM and implies that there is something essentially correct about MOND. And yet, the dark matter paradigm works very well on a cosmic scale: it provides an explanation of the acoustic oscillations evidenced by the anisotropies in the CMB and in the large-scale distribution of galaxies; it provides a consistent explanation of the formation of structure; and, in combination with dark energy, it explains the observed form of the Hubble diagram of distant supernovae. MOND in its present incomplete form does not address these issues. So the question becomes – how can we reconcile the large-scale success of ΛCDM with the galaxy-scale predictive power of MOND? This surely is one of the most profound puzzles of modern cosmology.

9

Dark Matter, MOND and Cosmology

9.1 The Puzzle

Dark matter does not have the predictive power of MOND on the scale of galaxies. Yet cold dark matter has gained almost unquestioned acceptance by the majority of the relevant communities, while MOND has languished for more than 30 years outside of the mainstream. This is not because of some grand conspiracy against the idea of modified dynamics, but rather because of strong social factors that maintain support for the prevailing paradigm. There is an overriding tendency for scientists to function within the established framework and to select data that reinforce rather than challenge it (an effect that is supported by competition for academic positions and grants). There is a very large community (thousands) of physicists, astronomers and cosmologists with vested career interests in searching for and detecting dark matter – underground, above ground, in space, under-water, in the Antarctic ice. Moreover, in this case there is also an understandable reluctance to tamper with the historically established laws of physics; this is not what astronomers do. Physicists have a different culture, but most models for dark matter involve extensions of the standard model of particle physics; this sort of new physics is of general interest to the relatively large community of high-energy physicists and is perceived as less intrusive than modifying the venerable laws of Newton.

Beyond social factors, though, there is a reductionist current in modern science in general that in this area of research assigns priority to cosmology over mere galaxy phenomenology. Although the respectability of cosmology is quite recent, the science of the entire Universe is now considered to be more fundamental than that of its individual constituents. The pattern of anisotropies in the CMB is very well explained by the standard cosmological model, albeit with a somewhat unnatural combination of six free parameters. And given the precision of the fit to the angular power spectrum of these anisotropies, then, the reasoning goes, the

115

theory must be correct even in its perceived implications for galaxies. So most cosmologists tend to be dismissive of mere galaxy phenomenology and its wealth of regularities; these are details and are due to messy baryonic physics that will be understood someday in the context of more detailed computations of the processes of star formation and feedback. It is certainly not serious enough to require a modification of Newton's (and therefore Einstein's) laws.

Support for MOND tends to be concentrated in the community of observers and theorists who work on galaxies. Those scientists who take the idea seriously are impressed by the predictive power of MOND with respect to galaxy phenomenology: first of all, by the fact that every rotation curve is, in a real sense, a prediction and not a fit. This is particularly true with respect to low-surface-brightness, gas-dominated galaxies in the deep MOND limit, where the form of the interpolating function between MOND and Newton becomes irrelevant and the mass-to-light ratio of the stellar component is no longer a free parameter. And secondly, by the fact that the Tully–Fisher relation, the strongest correlation observed in extragalactic astronomy, follows naturally from such an acceleration-based modification of gravity or dynamics as an aspect of physical law and not the individual random dynamical histories of galaxies. Finally, there is the ubiquity of the critical acceleration, a_0, as the acceleration below which the discrepancy appears in galaxies, as the normalization of the Tully–Fisher relation, as defining a critical surface density or brightness that marks the transition from systems with and without a significant discrepancy. The fact that $a_0 \approx cH_0$ provides an apparent cosmological significance to this parameter. That there is not a generally accepted relativistic theory is seen, in this community, as a problem for the future.[1]

And yet, it is difficult to deny the evidence in support of CDM – or something very much like it – on cosmic length, and timescales. The primary evidence is provided by the angular power spectrum of the anisotropies in the cosmic background radiation. The relative amplitudes of the acoustic peaks are entirely consistent with the oscillation of the baryon–photon fluid in the pre-existing rigid potential wells of non-interacting dark matter. The implied abundance of the dark matter is about 25%, which is consistent with the Hubble diagram of distant supernovae: 30% non-relativistic matter (including 5% baryons) and 70% dark energy, causing the accelerated expansion of the Universe.

So what is the solution to this puzzle? Is there dark matter on a cosmic scale but modified dynamics on a galactic scale? Has the dark matter decayed since the epoch of decoupling of photons and baryons? Are there particles that can cluster on the scale of the acoustic fluctuations and down to galaxy clusters but not on the scale of galaxies? Is there a single field that behaves like dark matter on a large scale but modified dynamics on a smaller scale? Can the phenomenology of dark matter, dark energy and MOND be unified in a single theory?

There are two diametrically opposed possibilities: (1) particle dark matter does exist but cannot manifest its presence on the scale of galaxies; (2) there is no dark matter and the cosmic mass discrepancy is an aspect of the MOND (or its relativistic extension) on a large scale.

9.2 Particle Cosmic Dark Matter

9.2.1 Neutrinos

Can dark matter and MOND both be correct? Empirically, dark matter works well on large scales – from galaxy clusters (co-moving scales of 30–60 million light years) to the first few peaks in the power spectrum of the CMB anisotropies (100–600 million light years) up to the Hubble radius (13 billion light years); whereas, in view of the phenomenology so well encapsulated by MOND, dark matter fails to play any role in galaxy dynamics. Therefore, an obvious possibility is for dark matter to be in the form of particles that can collapse and cluster on a large scale but not on a small scale.

Neutrinos are obvious candidates because they are known to exist; they are stable, electrically neutral, and weakly interacting with baryons. A background of cosmic neutrinos exists, roughly equal in number density to the CMB photons, and it is now known that neutrinos have a small but finite mass. So it is accurate to say that non-baryonic particle dark matter certainly does exist in the form of neutrinos, although the contribution of this component to the total mass–energy density of the Universe remains uncertain. This is because the actual mass of the various neutrino types is unknown.

There are three types or flavors of standard neutrinos (electron, muon and tau) corresponding to the three generations of fermionic matter. A recent major result of experimental physics is that the three neutrino types can change into one another – they oscillate. This can happen only if these three types have non-zero mass.[2] The oscillation experiments tell us the differences between the masses of various neutrino types (in fact the square of the mass differences), but not the actual masses, and therefore provide a lower limit – about 0.05 eV – on the mass of a given flavor. If the actual mass were near this lower limit, then neutrinos would not make a cosmologically significant contribution to the energy density of the Universe ($\Omega_\nu < 0.001$). But the mass of the neutrino types could be larger. An experimental upper limit to the mass of the electron neutrino is about 2 eV (from direct β-decay experiments). If the mass were near this limit then, given the small mass differences between the types, this would be the actual mass of all three types, and the contribution to the mass–energy density of the Universe would be significant – at $\Omega_\nu \approx 0.1$, twice the contribution of baryons.

Neutrinos stop interacting with the rest of the primordial hot soup very early, when the average energy of particles and photons is several million electron volts (when the Universe is a fraction of a second old). Thus normal neutrinos would be highly relativistic when they decouple and appear as a separate component; the neutrino fluid would fall into the category of *hot* dark matter. Because the neutrinos are light and fast, fluctuations would first collapse into large-scale structure (clusters of clusters) rather than into normal galaxies.

But in fact, neutrinos with mass less than 2 eV could never form galaxy halos: neutrinos with spin-1/2 are fermions; the Fermi exclusion principle applies to such particles, which means that a maximum of only two particles can be packed into a single tiny cell in a position–momentum phase space (with volume equal to Planck's constant cubed). This means that there is a packing pressure that restricts the density of systems with a given spread in velocity. If the mass of the neutrinos is as low as the experimental upper limit of 2 eV, they could accumulate on the scale of clusters of galaxies[3] but not in individual galaxies because of the packing pressure. This would appear to be exactly what we require of dark matter in the context of MOND – a dark matter component of clusters of galaxies but not in individual galaxies where MOND explains the discrepancy.

There is an observational problem with this idea. Assuming that there is no effect of MOND at the epoch of decoupling, we can calculate the expected angular power spectrum of the CMB anisotropies when the dark matter is in the form of 2-eV neutrinos. This predicted power spectrum does not match the observed pattern of acoustic oscillations. In particular, the relative height of the third peak in the angular power spectrum is too low. This is because, with only 2-eV neutrinos, the rigid potential well of the pre-existing halo is not deep enough to provide the observed fluctuation on this scale.

A way around this problem is to consider a more massive neutrino – not one of the standard three but an additional, hypothetical *sterile* neutrino that does not interact via any of the standard-model forces except gravity (such particles are said to be theoretically well-motivated). Gary Angus pointed out that if the mass of this sterile neutrino is about 11 eV then it can make up the entire budget of dark matter in the Universe, as implied by the CMB anisotropies, and yet it still does not cluster significantly on the scale of galaxies. Because, on a large scale, such a neutrino effectively behaves like CDM, it produces a reasonable fit to the angular power spectrum of anisotropies – at least it did until the *Planck* satellite results were released.

Planck measures anisotropies down to a co-moving scale of 100 million light years (fourth to fifth acoustic peak). On this scale, 11-eV neutrinos do not behave like CDM because free streaming of the particles when they are relativistic damps the strength of the perturbations. Thus the fit to those anisotropies that result from these smaller-scale perturbations is not so perfect as that of pure CDM.

In addition to these empirical difficulties in fitting the very high harmonics of the acoustic oscillations, the model of neutrinos plus MOND appears contrived. Two independent constructions are required to separately explain the cosmological and galaxy phenomenology; it is not an aesthetic universe.

9.2.2 Soft Bosons

Bosons are particles with integral spin (0, 1, 2,...) and as such there is no constraint on their density; unlike fermions, all of the particles comprising a halo can be found in a single quantum state (a Bose–Einstein condensate). However, there is a constraint on the size of the region they can occupy which is set by the de Broglie wavelength, the quantum uncertainty in the position of a particle; the bosons cannot be located within a scale given by $\lambda \approx \hbar/(mv)$, where v is the velocity. If the mass m is very small (less than 10^{-27} eV) then that scale can be large – considerably larger than an entire galaxy if the spread in velocity is on the order of the internal velocity of a galaxy (30 million light years if ≈ 200 km/s). For this reason they have been called "soft bosons." Such particles, although cold, could not form galaxy halos.[4]

A cosmic scalar field with an associated potential energy can behave as dark energy and dark matter (Chapter 6). If the field, ϕ, has a quadratic potential $V(\phi) \approx m^2\phi^2$, then its value may oscillate about the minimum of that potential ($\phi = 0$ in this case); these oscillations behave as pressureless dark matter (cold dark matter) particles with mass m and, having spin zero, constitute bosons. The period of these oscillations is h/mc^2; when the age of the Universe is shorter than this period, the field is frozen onto the side of the potential wall (position 1 in Figure 6.3) and constitutes dark energy. In the example given, the dark energy phase would be on the order of 100 000 years, or shortly before decoupling of photons from baryons. But then, when the Universe becomes older than this timescale, the field rolls down the well and begins to oscillate about the minimum of the potential well – effectively forming scalar particles that behave as cold dark matter. This dark matter is born spontaneously and was never in thermal equilibrium with the rest of the Universe.

With a mass less than 10^{-27} eV, such long-wavelength bosons could not form galaxy halos, but they could accumulate on the scale of galaxy clusters as well as that of the acoustic waves in the primordial baryon–photon fluid – those sound waves that are observed in the CMB anisotropies. So these bosons could satisfy conditions necessary for cosmological dark matter while not contributing to the mass discrepancy in systems smaller than clusters of galaxies.

However, could such a model also result in modified dynamics? Most of the relativistic extensions of MOND involve at least one additional scalar field, but it is possible that there are two scalar fields – a bi-scalar theory. The first carries the additional acceleration-dependent MOND force, and the second effectively represents the interpolating function between the MOND and Newton limits. It is the second field that we would identify as the oscillating field – the field creating the bosons. So it is possible to construct a theory that includes MOND, cosmic dark matter and dark energy.[5]

There are variations on this theme. Lasha Berezhiani and Justin Khoury have suggested that, with bosons having mass on the order of an electron volt, it is possible to construct a model of galactic halos in which the bosons merge into the lowest quantum state – they become a Bose–Einstein condensate. This halo is a superfluid and waves in the superfluid can be constructed to mediate a MOND-like force between the halo and the baryons. In galaxy clusters, with velocities higher random velocity, the bosons behave as usual CDM particles – thus explaining MOND phenomenology on galaxy scales but dark matter on cluster and larger scales.[6]

9.3 New Physics

Milgrom has proposed a relativistic *bimetric* theory of MOND.[7] In this theory, there are two different space–times and we live in one of them. The gravity is modified due to the interaction of the two space–times – an interaction that involves a term that is dependent upon gravitational accelerations. There is one new constant, a_0, and the theory approaches pure general relativity as a_0 vanishes. Both MOND and dark energy arise from this same interaction term and the scale of both effects is given by a_0. Matter may be present in the second space–time – "twin matter" – which may or may not directly affect our space–time, but it does open up the possibility of perfect dark matter – undetectable apart from its gravitational interaction.

The theory is suggestive of so-called "brane-world" models in which there is a universe parallel to that of our own – a universe that we cannot directly experience but which has a dynamical effect on our Universe that, in this case, is observable as modified Newtonian dynamics at low accelerations.[8] The idea, so far, has a number of loose ends – too many to derive a definite cosmology – but there appear to be many possibilities for simulating the effects of cosmic dark matter.

One of these possibilities is that suggested by Laura Bernard and Luc Blanchet.[9] Here, each of the two space–times has its own special sort of dark matter; in particular, our space–time contains a dipolar dark matter, characterized by a gravitational dipolar moment that is able to polarize this medium in the presence of the gravitational field of ordinary matter (this would be analogous to the polarization of a dielectric medium in the presence of an imposed electric field). In this case, the polarization enhances the gravitational field in the limit of low accelerations. In the context of this theory, the auxiliary space–time also contains dark matter that can interact with the dipolar dark matter in our space–time. The authors argue that the theory can reproduce the phenomenology of MOND on a small scale and ΛCDM on a large scale. It does, however, require several additional constructions: two space–times with two different forms of dark matter that interact via a particular form of force; in addition, the dark matter in our world must be of a hypothetical dipolar sort.

A different speculative idea involves the possibility of preferred frames. Isaac Newton thought in terms of absolute space and time – constructions that exist independently of matter and act as an inert stage on which the actors, the dynamics of matter and gravity, play out their roles. Newton recognized the equivalence of inertial frames but thought that there existed one such frame that was really special – that of the absolute space. He also codified the concept of universal time that defines simultaneity in different locations and flows by itself independently of any external influence. Now, we all know that Einstein overturned these ideas by proposing that the speed of light is the same in all inertial frames; that there is no special frame but that time flows differently in different frames in relative motion; that the very concept of simultaneity should be abandoned.

The question of a preferred frame arises now in a cosmological context because we can clearly identify, operationally, a special universal frame – that which is at rest with respect to the cosmic background radiation. In the context of the FLRW cosmology, there is also a special time – cosmic time – that can be constructed in the context of the assumed isotropy of space, the Cosmological Principle, and effectively restores the concept of simultaneity throughout the Universe.

Does the Universe as a whole really possess Einstein's equivalence of moving frames of reference? Or is there actually a preferred frame where the laws of physics take their simplest form? There are modern relativistic theories involving additional vector fields that generically violate the equivalence of frames with respect to gravitational phenomena, hence they are typically called Einstein–aether theories[10] (Bekenstein's TeVeS is such a preferred-frame theory). Such theories give rise to observable aether drift effects, such as a diurnal variation in the constant of gravity. The fact that these effects are not observed to high precision means that any such dynamical effects of a preferred universal frame must be strongly suppressed on the scale of the Solar System.

The simplest such theory with a preferred frame involves a single new scalar field. This field has been called the "khronon" after the ancient Greek word for sequential time, because the khronon is, in effect, the time coordinate in the universal frame, and vectors perpendicular and parallel to a surface of constant khronon field (constant cosmic time) allow the theory to be written as an Einstein–aether theory (essentially, the vector perpendicular to this surface defines the direction of cosmic time). The motivation for this idea has nothing to do with MOND but is an attempt to construct a theory joining general relativity with quantum mechanics – a theory of quantum gravity.[11] But a modification of the khronon theory, a modification yielding MOND, allows the effects of a universal preferred frame to be hidden locally.[12] Basically, a gravitational acceleration greater than a_0 suppresses aether drift effects locally (a_0 also defines the scale of a cosmological constant). MOND appears in the transition between preferred-frame cosmology and frame-independent local dynamics. So far, it is not clear how the theory can confront cosmological observations such as the CMB anisotropies and the observed structure in the Universe.

This is also true of another recent speculation that is built around the ideas of the string theorist Erik Verlinde.[13] The idea is that Newtonian attraction is an entropic force. An entropic force is not a fundamental force such as the electromagnetic force mediated by photons, but is a force that results from the tendency of a system to become more random – to increase its entropy. For example, the tendency of a long polymer molecule to become bent and kinked (a more probable configuration) is an entropic force; osmotic pressure, the flow of water across a membrane to increase dilution of a solution, is due to an entropic force. In the case of gravity, Verlinde argues, the entropic gravity force emerges from microscopic bits of information stored on a holographic screen; in the presence of this screen, particles move in such a way as to maximize the randomness (entropy) on the hologram.

This is a fundamentally different view of the gravitational force and has generated much controversy.[14] However, when a minimum temperature for the screen is included – a temperature corresponding to that of the de Sitter horizon – then the gravitational force is modified into a form corresponding to MOND.[15] This essentially reproduces the result given in Section 8.3 and provides a natural identification of the MOND acceleration a_0 with the cosmological constant. It is not yet a theory but a heuristic derivation and the cosmological consequences have not been fully considered.

9.4 Reflections

There is no shortage of ideas for a cosmological origin of MOND or MOND combined with dark matter. In most of the scenarios described here, it is assumed that MOND has no dynamical effect at the epoch of decoupling. This assumption

9.4 Reflections

leads to the conclusion that there must be something very much like CDM at the point in time when the oscillations in the baryon–photon fluid are frozen into the cosmic background radiation. However, the unmodified nature of physics at the epoch of decoupling is very much an assumption and depends upon how a_0 evolves with time. For example, if the critical acceleration is in fact proportional to cH, not cH_0 (that is, the critical acceleration increases going back into the past as does H), then this parameter will be more than 10 000 times larger at decoupling. If so, the peculiar accelerations relevant to the acoustic oscillations will fall into the MOND regime. Then we might expect the development of perturbations not to be described by general relativity but by the appropriate relativistic extension of MOND.

Even if a_0 does not vary with cosmic time, the oscillations smaller than those corresponding to the second or third peak in the power spectrum will be, mildly, in the MOND regime. Thus it is unclear whether an additional dark matter component is necessary to explain the observations of the CMB anisotropies. But it is also unclear whether a completely non-standard theory would be able to match the very detailed cosmological observations with an efficiency comparable to ΛCDM. For example, Bekenstein's TeVeS, for relevant ranges of the free parameters of the theory, do provide a higher growth rate of perturbations (due to the energy density in the additional fields.[16] But with the standard assumptions (for example, the form of the primordial fluctuation spectrum) the theory does not match the power spectrum of the CMB anisotropies, even with the addition of 2-eV standard neutrinos,[17] nor the distribution of clusters by mass with the addition of 11-eV sterile neutrinos.[18]

With MOND, the basic gravitational or dynamical framework for considering cosmological phenomenology is unsettled. The fundamental field that is present in all theories is the space–time metric field of general relativity, but most of the suggested theories include additional fields that provide an unconventional force that becomes dominant at low accelerations. It is unclear what the essential new parameters are, but in any case the critical acceleration a_0 should be included, either as a fundamental constant or identified with an additional existent parameter (e.g., the cosmological constant, the Hubble radius); this is an empirical imperative. Several of the suggested theories (e.g., bimetric MOND, Einstein–aether theories) become equivalent to general relativity when a_0 is set to zero, which would seem to be a desirable attribute. Moreover, in these theories, a_0 is directly identified with the cosmological constant.[19]

In the context of ΛCDM, the basic gravitational framework is also provided by general relativity – a well-established theory with no adjustable parameters and no detected empirical contradictions on sub-galactic scales. In the cosmological context there are no unconventional forces acting on matter with the strength of gravity.

However, general relativity by itself cannot account for the details of cosmological observations; separate constructions must be introduced primarily as non-standard sources for the gravitational field. The most conspicuous of these is a cosmological constant or an evolving dark energy that causes the observed present accelerated expansion of the Universe and may be due to a cosmic field. If the inflationary scenario is the correct solution to the problem of special initial conditions, then there must also be a cosmic field that drives the exponential expansion of the very early Universe toward isotropy and flatness. The microphysics of these two additional fields is unknown. Although there are a number of speculations, these two constructs are added in an ad hoc way to achieve the desired empirical results.[20]

The same is true of the second "aether" of ΛCDM – the cold dark matter. Its abundance and properties are chosen to explain observed phenomena ranging from the anisotropies in the CMB – due supposedly to acoustic oscillations of the pre-decoupling baryon–photon fluid – down to the rotation curves of spiral galaxies. In this more local role it fails. But apart from the empirical problem presented by galaxy phenomenology, the nature of this major component is as uncertain as that of dark energy. Certain particle candidates (supersymmetric superpartners) are said to be "well motivated." This means that there is a class of hypothetical particles that emerge in the context of the best-guess extension to the standard model of particle physics. Even if such particles exist, they do not necessarily have the properties required to constitute the putative cosmic dark matter (the mass, the abundance). The ratio of the abundances of these three major components – dark energy to dark matter to baryons – is also ad hoc and is completely empirically motivated.

Nonetheless, the standard paradigm at present fares better on a cosmological scale than does MOND. This is because the theory, with these several attributes added to general relativity, can consistently explain the essential cosmological observations: the angular power spectrum of the anisotropies in the CMB; the large-scale distribution of observable matter and, in particular, the presence of the same baryon acoustic scale indicated by the CMB anisotropy; and the current accelerated expansion of the Universe revealed by the Hubble diagram of supernovae. However, the successes are more explanatory than predictive.[21]

MOND does not yet have a generally accepted relativistic extension but it has reached the point where the existing theories can begin to address these cosmological issues. MOND is epistemologically superior in its non-relativistic manifestation as a modification of Newtonian dynamics. Here it is predictive and successfully addresses the systematic and detailed phenomenology of the discrepancy in galaxies, while ΛCDM fails on this scale – it is neither predictive nor explanatory in spite of ad hoc attempts at semi-analytic repairs.

ΛCDM is clearly incomplete, and it is premature to make triumphal claims that the Universe is understood. The mood of triumphalism is at odds with a basic requirement of the scientific method: a continual *critical* re-examination and testing of the fundamental assumptions of a theory or paradigm. Even though enormous progress has been made in the last two decades, cosmology is not frozen and remains as lively and exciting as it has been for the past century. New ideas appear every day; most are not correct, but such speculation is in the nature of science. In this process we should always keep in mind that the study of the Universe as a whole also includes its smaller constituents as well as us – observing and thinking.

10
Plato's Cave Revisited

Suppose that Socrates and Glaucon were to wake up and continue their dialogue on the nature of reality and the acquisition of knowledge, all in the light of modern scientific developments. Perhaps it would go something like this:

"Well, Glaucon, you must remember our previous discussion concerning the illusionary nature of the physical world and the attainment of true knowledge of the eternal Forms."

"Yes, indeed I do – your Allegory of the Cave is truly unforgettable. Shall I summarize it for you?"

"Please do, Glaucon, so that we can perhaps re-interpret the metaphor after all these many years. In particular, given the predictive success of modern science and the resulting fascination with the physical world, we might discuss whether or not the allegory still contains a message for us."

"As I recall, it goes like this: A group of prisoners sits in a cave; they are constrained to face a nearby wall and look at nothing else, and they have been in this position for all of their cogent lives. Behind them and above there is fire that dimly illuminates the cave, and between the fire and the prisoners there is a road on which puppeteers, hidden by a low wall, pass carrying various objects. Some of the objects are images of humans and animals; some are more abstract. The puppeteers are free to speak and their voices are reflected off the wall. The objects that they carry cast shadows on the wall, and the moving shadows and reflected voices are all that the prisoners can see or hear of the world; this forms their view of reality. The prisoners are permitted to speak with one another and discuss what they see. Of course, they assume that the images before them on the wall are the real world since this is all they know of it."

"Excellent, Glaucon. Your memory is pretty good after 2500 years. And what happens next in the story?"

"We suppose that one of the prisoners is set loose and is forced to turn around – that is, away from the wall and into a position such that he can see the fire and what is happening behind the prisoners. His eyes are somewhat dazzled by the light of the fire and, at first, he cannot get a clear view of the forms passing along the road. But then, as his eyes adjust he begins to see that there is a deeper reality behind the shadows on the wall: there is the fire and the moving objects that cause the shadows. Now under the road, there is a tunnel that leads up to the exit from the cave and the world outside. The unchained prisoner is forced through this tunnel and up into the world of sunlight outside the cave – a world that he never knew existed."

"That must have been quite a shattering experience for our poor prisoner – to be suddenly exposed to the light of day."

"Yes, indeed. It would take quite a while for his eyes to become accustomed to the sunlight. He would first have to look at the shadows and reflections of the people and animals who inhabit this outer world, and of course he would be able to see the stars and moon at night. He would certainly conclude that these entities are more real than the shadows he saw previously in the cave. But finally he would be able to see the Sun itself in the full light of day. He would realize that this was the cause of the changing seasons, the growing plants, everything else that he could see around him, and everything that he and his fellow prisoners could see and hear in the cave. It is the ultimate reality – the Truth."

"And would he wish to return to the cave and tell his fellow prisoners what he had experienced?"

"At first, no. Because he would be happy to be in the world above, at a more profound and complete level of reality. He would have no desire to return to the dark cave where people lived with the illusions of the shadow world. But finally, out of a grudging sense of obligation to his unfortunate fellow prisoners, he would return to the cave to enlighten his former companions."

"And that would surely be a disconcerting experience for the wayward prisoner."

"Yes indeed. At first his eyes, accustomed to the light, would not discern the dim surroundings of his previous life. He would appear awkward and out of place. The prisoners, seeing his discomfiture, would laugh at him and tease him for being so bold as to venture out of their safe environment."

"And I suppose the prisoners would have developed their own system of knowledge and of explaining the world that they see and hear."

"This is true. They would have developed their own science, and some of the prisoners, being certainly very clever, would have received prizes and awards for creating models and theories to explain the shadow world of their perceptions."

"And Glaucon, do you think that these prisoners would be happy to see the returned inmate with a very different view of the world and hear his story of another level of reality lying beyond that of their experience?"

"Certainly not, Socrates. They would be quite upset with the returner, and some of them might even wish to do him harm."

"And now can you tell me the meaning of this allegory?"

"The meaning is that if we live only in the world that is apparent to our senses – a physical world – then we live in the world of the cave where all is transient, not in a world of the Forms that are eternal and unchanging. Beyond the shadows and sounds that we see and hear there is a deeper reality – a cause for these shadows that is not immediately obvious. And beyond that there is a higher and brighter truth, and that is Truth itself – pure and abstract."

"You have learned well Glaucon. But now it is more than two millennia later, and the new philosophy of empiricism has developed, with a different criterion for truth and a new methodology called science. Humans through science have discovered much about the physical world, and, significantly, about the rationality behind that physical world. Do you not think that the meaning of our story might have changed after all this time and all of these developments?"

"I suppose it must have, Socrates, but I have been doing most of the talking this time. Perhaps you can tell me in what ways it has changed. Are not the eternal Forms, after all, eternal? Is not the reality that is apparent to our senses and to those instruments that extend our senses only, at best, an imperfect reflection of those Forms?"

"The story of the cave has different levels, Glaucon, and over the years it has been used by many people for many purposes. But I believe that the essential message remains: the shadows on the wall of the cave are not the ultimate reality but are a reflection of that reality. Now humans in their cleverness have observed the oldest light in the world on the wall of the sky – the cosmic microwave background radiation – and they have measured tiny fluctuations in that light on very small scales. But just like the shadows on the wall of the cave, those patterns in the background light are but a trace of the ultimate reality. The patterns require theory and models for interpretation and understanding, and cosmologists attempt to provide that understanding. But those who stare only at these patterns on the wall and do not turn around to look at other phenomena will never grasp the full picture of the world. While the criteria for judging our theories can only be empirical, beyond that, it is the case that those theories that most perfectly reflect the world of our senses, the ones that approach the truth, are also those that are the most aesthetic and beautiful; they have a simplicity – without extra attributes or unreal aethers – and reflect the underlying rationality and mathematical perfection of the world. This is truly amazing – this and the fact that we can begin to comprehend this perfection with our limited minds, and from a small corner of the Universe is the brilliant Sun shining upon us all – the ultimate Beauty and the ultimate Truth."

"Thank you, Socrates. I look forward to having this dialogue again someday, and hopefully we will not have to wait so long for the next time."

Notes

Introduction

1. The problem of falsifying the existence of dark matter and similar issues are discussed in the conclusion of my book *The Dark Matter Problem: A Historical Perspective*, Cambridge: Cambridge University Press, 2010.
2. See E. Bianchi and C. Rovelli (2010). Why all these prejudices against a constant? (arXiv:1002.3966) for a highly readable discussion of these issues. Here the authors argue that the coincidence problem is often demonstrated in terms of time plotted logarithmically rather than linearly and that this gives a false impression of the range of coincidence. This is true, but in the context of a naturalistic explanation (as opposed to anthropic), the problem remains because at some early epoch the magnitude of the cosmological constant must be carefully tuned to the present matter density in order to promote this commensurability over a large range of cosmic time.
3. Milgrom published three papers in 1983 that presented the essence of the algorithm and its predictions for galaxies and galaxy systems (M. Milgrom (1983). *Astrophys. J.* **270**, 365; 371; 384). Many of these predictions have subsequently been verified.

1 Creation Mythology

1. L. W. King (2002). *Enuma Elish: The Seven Tablets of Creation*, reprint. Available online at http://king-of-heroes.co.uk. Names were important in ancient mythology. If a god or a natural manifestation of the world did not bear a name, it did not exist.
2. G. Smith, *The Chaldean Account of Genesis*, London: Scribner, Armstrong & Company, 1876. Of course there are significant differences, given the Hebrew emphasis of a single deity that stands apart from and controls nature.
3. E. Hamilton, *Mythology*, New York: Little, Brown, 2013. It is interesting that in Greek mythology, the creation of the world is not explained as the act of any particular god; it happens spontaneously.
4. The first cause would more properly belong to the realm of traditional metaphysics, although there are current speculative theories of pre-Big Bang cosmology.
5. Hindu cosmology is complicated, with several traditions being expressed in various texts. The idea of an eternal universe with cycles of creation and destruction is derived from the ancient Vedic tradition but with elaborations in the Puranic texts – particularly considering questions of timescale.
6. See the influential essay by P. W. Anderson (1972). More is different, *Science* 177, **393** on emergence vs. reductionism in physics. A more extended discussion is given by Robert Laughlin in his book, *A Different Universe: Reinventing Physics from the Bottom Down*, New York: Basic Books, 2005.

7. An early application and definition of the term "anthropic principle" appeared in the influential contribution of Brandon Carter: "Large number coincidences and the anthropic principle in cosmology" in M. S. Longair (ed.) *Confrontation of Cosmological Theories with Observational Data, Proc. IAU Symposium* 63, pp. 291–298, Amsterdam: Springer, 1974.
8. If the world is a multiverse in which each separate universe has its own physics, with different physical constants and, perhaps, different dimensionality, and if there are an infinite number of such universes, then, in a real sense, everything is predicted. If everything is predicted then nothing is predicted.

2 Three Predictions of Physical Cosmology

1. This is an idealized description of the scientific method as perceived by a practitioner, not by a more objective philosopher of science. There has been much discussion over several centuries on the role of induction vs. deduction in science, but it is clear that science is much more than the random collection of observations and facts; it is systemized by theory, the value of which is measured by often unexpected, successful predictions. It is also clear that, in spite of the ideal of objectivity, scientists come complete with a set of prejudices about the way the world works, or should work, and, as emphasized by Thomas Kuhn, these preconceptions are set by the prevailing paradigm – the dominant framework within which young scientists are educated.
2. The correspondence between Bentley and Newton is described in J. North, *Cosmos: An Illustrated History of Astronomy and Cosmology*, Chicago, IL: University of Chicago Press, 2008.
3. J. North, *The Measure of the Universe*, Oxford: Oxford University Press, 1965.
4. J. A. Wheeler and K. W. Ford, *Geons, Black Holes, and Quantum Foam: A Life in Physics*, New York: W. W. Norton & Company, 2000.
5. The repulsive force due to the cosmological term is given by Λr, where Λ is the value of the cosmological term and r is the distance from any given point. This must balance the attractive gravity force in a uniform homogeneous medium with a constant density, ρ; that is, $4\pi G\rho/3$. Equating the repulsive and attractive forces, we find that this is only possible if the density of the medium is given by $\rho = 3\Lambda/(4\pi G)$.
6. The present value of the critical density depends only upon the present expansion rate and is given by

$$\rho_c = \frac{3H_0^2}{8\pi G}$$

where G is the gravitational constant. With an $H_0 \approx 70$ km/s per Mpc (Mpc is the abbreviation for megaparsecs or one million parsecs and is equal to 3.26 million light years, a typical distance unit in extragalactic astronomy). The density parameter Ω is the density of the Universe in terms of the critical density, $\Omega = \rho/\rho_c$. Ω may be broken down into the separate constituents of the Universe: that in pressure free matter, Ω_m, in radiation, Ω_r, in baryons, Ω_b, or in vacuum energy, Ω_v.
7. See H. Nussbaumer and L. Bieri, *Discovering the Expanding Universe*, Cambridge: Cambridge University Press, 2009, for a very complete discussion of these developments.
8. E. Hubble (1929). A relation between distance and radial velocity among extra-galactic nebulae, *Proc. Natl. Acad. Sci. USA*, **15**(3), pp. 168–173.
9. The Hubble constant is typically given in units of kilometers per second per megaparsec, where a megaparsec is an astronomical unit of distance relevant to extragalactic scales. It is 3.26 million light years. These units are equivalent to inverse time and, in the most recent determinations, would be $(14 \text{ billion years})^{-1}$.
10. The space–time metric is a mathematical object – a tensor – that determines the geometric properties of space and time (the curvature, for example). In four-dimensional space it has 16 components but because of the symmetry of the tensor only ten components are independent. The metric also appears in the calculation of intervals between points, or events, in the space time; this interval is an invariant quantity; which is to say, the interval does not depend upon the

particular coordinate system used to label points in space–time. In Einstein's theory, the metric replaces the Newtonian gravitational potential.
11. The early dominance of radiation was first realized by Richard Tolman in *Relativity, Thermodynamics, and Cosmology*, Oxford: Clarendon Press, 1934.
12. C. F. von Weizsäcker (1937). Über Elementumwandlungen im Innern der Sterne. I (On transformations of elements in the interiors of stars. I), *Physikalische Zeitschrift* **38**, 176–191; and H. A. Bethe (1939). Energy production in stars, *Phys. Rev.* **55**, 434–456.
13. The Boltzmann equation applies in the condition of thermodynamic equilibrium. It relates the abundances of chemical or atomic species to their energy difference and the temperature of the surrounding medium.
14. The many contributions of George Gamow are reviewed in H. Kragh (1996). Gamow's game: The road to the hot Big Bang, *Centaurus*, **38**, pp. 335–361.
15. A highly readable and entertaining account of the clash between the Big Bang and Steady State world views can be found in S. Singh, *Big Bang*, New York: Harper Perennial, 2005.
16. The first chapter of Schwarzschild's classic book, *Structure and Evolution of the Stars*, Mineola, NY: Dover, 1958, provides an excellent introduction to the study of stellar evolution.
17. E. M. Burbidge, G. R. Burbidge, W. A. Fowler, and F. Hoyle (1957). Synthesis of the elements in stars, *Rev. Mod. Phys.* **29**, 547–650.
18. F. Hoyle and R. J. Taylor (1964). The mystery of the Cosmic helium abundance, *Nature* **203**, 1108–1110.
19. The discovery and early observations of the CMB are described in P. J. E. Peebles, L. A. Page, Jr., and R. B. Partridge (eds.), *Finding the Big Bang*, Cambridge: Cambridge University Press, 2009.
20. D. J. Fixsen, E. S. Cheng, J. M. Gales, J. C. Mather, R. A. Shafer, and E. L. Wright (1996). The cosmic microwave background spectrum from the full COBE FIRAS data set, *Astrophys. J.* **473**, 576–587.
21. For small fluctuations the force of gas pressure is sufficient to prevent gravitational collapse. For large fluctuations the gravity force overwhelms the pressure force and collapse occurs.
22. R. K. Sachs, A. M. Wolfe (1967). Perturbations of a cosmological model and angular variations of the microwave background, *Astrophys. J.* **147**, 73–90.
23. G. F. Smoot, et al. (1992). Structure in the COBE differential microwave radiometer first-year maps, *Astrophys. J.* **396**, L1–L5; and C. L. Bennett, et al. (2003). First-year Wilkinson microwave anisotropy probe (WMAP) observations: Preliminary maps and basic results, *Astrophys. J. Supp.* **148**, 1–27.

3 The Very Early Universe: Inflation

1. R. H. Dicke, *Gravitation and the Universe*, Philadelphia, PA: American Philosophical Society, 1970.
2. The origins of the idea of inflation, particularly the fundamental contribution of Soviet physicists, is described by C. Smeenk, "False vacuum: early Universe cosmology and the development of inflation" in A. J. Knox and J. Eisenstaedt (eds.) *The Universe of General Relativity*, pp. 223–257, Basel: Springer, 2002.
3. A. Guth (1981). Inflationary Universe: a possible solution for the horizon and flatness problems, *Phys. Rev. D* **23**, 347–356; A. Linde (1982). A new inflationary universe scenario: a possible solution of the horizon, flatness, homogeneity, isotropy, and primordial monopole problems, *Phys. Lett. B* **108**, 389–393.

4 Precision Cosmology

1. P. J. E. Peebles (1966). Primordial helium abundance and the primordial fireball. II, *Astrophys. J.* **146**, 542–552; R. V. Wagoner, W. A. Fowler, and F. Hoyle (1967). On the synthesis of elements at very high temperatures, *Astrophys. J.* **148**, 3–49.

2. The appropriate term "baryometer" was invented by David Schramm and Michael Turner (1998). Big-bang nucleosynthesis enters the precision era, *Rev. Mod. Phys.* **70**, 303–318.
3. The historical developments relating to the concept of dark matter as well as the relevant observations are discussed at some length in my book *The Dark Matter Problem: A Historical Perspective*, Cambridge: Cambridge University Press, 2010.
4. S. D. M. White, J. F. Navarro, A. E. Evrard, and C. S. Frenk (1993). The baryonic content of galaxy clusters: A challenge to cosmological orthodoxy, *Nature* **366**, 429–433. This is a seminal paper concerning the universal ratio of baryonic to dark matter leading, in no small degree, to the shift in paradigm between standard and ΛCDM.
5. A. Riess, et al. (1998). Observational evidence from supernovae for an accelerating universe and a cosmological constant, *Astron. J.* **116**, 1009–1038; S. Perlmutter, et al. (1997). Measurements of Omega and Lambda from 42 high-redshift supernovae, *Bull. Am. Astron. Soc.* **29**, 1351.
6. P. de Bernardis, et al. (2000). A flat universe from high-resolution maps of the cosmic microwave background radiation, *Nature* **404**, 955–959; and S. Hanany, et al. (2000). MAXIMA-1: A Measurement of the cosmic microwave background anisotropy on angular scales of 10'–5°, *Astrophys. J.* **545**, L5–L9.
7. C. L. Bennett, et al. (2003). First-year Wilkinson microwave anisotropy probe (WMAP) observations: Preliminary maps and basic results, *Astrophys. J. Supp.* **148**, 1–27.
8. P. A. R. Ade, et al. (2014). Planck 2013 results. I. Overview of products and scientific results, *Astron. Astrophys.* **571**, A1.
9. A very complete review of the physics of acoustic oscillations and a list of references up to 2002 is provided in W. Hu and S. Dodelson (2002). Cosmic microwave background anisotropies, *Ann. Rev. Astron. Astrophys.*, **40**, 171–216.
10. *Planck* Collaboration (2014). *Planck* 2013 results. XVI. Cosmological parameters, *Astron. Astrophys.* **571**, A16.
11. There are fluctuations of growing and declining amplitude, but inflation delivers growing adiabatic fluctuations and this provides the temporal phase coherence.
12. The significance of temporal phase coherence is described in detail in A. Albrecht, D. Coulson, P. Ferreira, and J. Magueijo (1996). Causality, randomness, and the microwave background, *Phys. Rev. Lett.* **76**, 1413–1416.

5 The Concordance Model

1. J. P. Ostriker and P. J. Steinhardt (1995). The observational case for a low-density Universe with a non-zero cosmological constant, *Nature* **377**, 600–602.
2. Much of this discussion may be found in D. Larson, J. L. Weinland, G. Hinshaw, and C. L. Bennett (2015). Comparing *Planck* and WMAP: Maps, spectra and parameters, *Astrophys. J.* **801**, 9.
3. The relation is $\eta_B = 274 \Omega_b h^2 \times 10^{-10}$, where h is the Hubble parameter in units of 100 km/s per Mpc. So determination of η_B is equivalent to determination of $\Omega_B h^2$. In the discussion here I take $H_0 = 70$ km/s per Mpc.
4. Some stars apparently produce lithium. The element can also be produced in the interstellar medium by collisions of heavier atoms with cosmic rays.
5. A. Coc, J.-P. Uzan, and E. Vangioni (2014). Standard big bang nucleosynthesis and primordial CNO abundances after *Planck*, *J. Cosmol. Astropart. Phys.*, **2014**(10), 050.
6. See G. Steigman (2010), Primordial nucleosynthesis: The predicted and observed abundances and their consequences (arXiv:1008.4756) for an excellent review circa 2010.
7. This statement actually requires some qualification. The maximum luminosity (the intrinsic power of the supernova) is dependent upon the rate at which the brightness in the supernova event declines; the longer the decay time, the more luminous the supernova. The true maximum luminosity must be determined by correcting for this effect that is calibrated through observations of local supernovae. Disturbingly, there is no clear understanding of the mechanism underlying this important correction.

8. For example, there is the supernovae legacy survey described in M. Sullivan, et al. (2011). SNLS3: Constraints on dark energy combining the supernova legacy survey three-year data with other probes, *Astrophys. J.* **737**, 102.
9. R. R. Caldwell (2002). A phantom menace? Cosmological consequences of a dark energy component with super-negative equation of state, *Phys. Lett. B* **545**, 23–29. Phantom dark energy is considered unphysical by many because it violates the dominant energy condition, thought to be necessary to protect the world from time machines and wormholes.
10. A. G. Riess, L. Macri, S. Casertano, H. Lampeitl, H. C. Ferguson, A. V. Filippenko, S. W. Jha, and R. Chornock (2011). A 3% solution: determination of the Hubble constant with the Hubble Space Telescope and Wide Field Camera 3, *Astrophys. J.* **730**, 119.
11. R. Wojtak, A. Knebe, W. A. Watson, I. T. Iliev, S. He, D. Rapetti, G. Yepes, and S. Gottl (2014). Cosmic variance of the local Hubble flow in large-scale cosmological simulations, *Mon. Not. R. Astron. Soc.* **438**, 1805–1812.
12. R. A. Daly, et al. (2008). Improved constraints on the acceleration history of the Universe and the properties of the dark energy, *Astrophys. J.* **677**, 1.
13. D. Eisenstein, W. Hu, and M. Tegmark (1998). Cosmic complementarity: H_0 and Ω_m from combining cosmic microwave background experiments and redshift surveys, *Astrophys. J.* **504**, L57–L60.
14. For example, see the Sloan Digital Sky Survey (SDSS): M. Strauss and G. Knapp (2005). The Sloan Digital Sky Survey, *Sky and Telescope* **109**, 34.
15. D. Eisenstein, et al. (2005). Detection of the baryon acoustic peak in the large-scale correlation function of SDSS luminous red galaxies, *Astrophys. J.* **633**, 560–574.
16. When correcting this motion of the Sun by its velocity with respect to the local group of galaxies, then it is found that the peculiar motion of the local group with respect to the CMB is 600 km/s.
17. This amusing moniker was first applied in K. Land and J. Magueijo (2005). The axis of evil, *Phys. Rev. Lett.* **95**, 071301.
18. For a review, see C. Copi, D. Huterer, D. Schwarz, and G. Starkman (2010). Large-angle anomales in the CMB, *Advances in Astronomy* **2010**, id. 847541.

6 Dark Energy

1. Here again, the must useful reference remains the old book by Martin Schwarzschild, *The Structure and Evolution of the Stars*, Mineola, NY: Dover, 1958.
2. See D. Branch (1998). *Ann. Rev. Astron. Astrophys.* **39**, 78, for a general discussion of the possible progenitors.
3. The dust would have to be different from that in the interstellar medium of galaxies. The usual dust particles, having sizes comparable to that of the wavelength of visible light, scatter preferentially blue light which reddens the color of the background object (as the Sun is reddened by dust in the Earth's atmosphere, particularly noticeable at sunrise or sunset). Dust grains that could dim but not redden, i.e., grey scattering, would have to be larger in size.
4. See S. Alexander, T. Biswas, A. Notari, and D. Vaid (2009). Local void vs dark energy: Confrontation with WMAP and type Ia supernovae, *J. Cosmol. Astropart. Phys.* **2009**(9), 025, for a recent discussion of the effects of a local void on the apparent expansion of the Universe.
5. See the review by J. A. Frieman, M. S. Turner, and D. Huterer (2008). *Ann. Rev. Astron. Astrophys.* **46**, 385, for a discussion of the direct and indirect evidence for an accelerating expansion as well as the theoretical aspects of dark energy.
6. Steven Weinberg first emphasized this issue in an important review paper: (1989). The cosmological constant program, *Rev. Mod. Phys.* **61**, 1. He suggested an anthropic solution to the problem. We could not exist in a universe with such a large vacuum energy density, so it must be largely removed by a negative cosmological constant. His guess for the resulting value is very close to that currently observed. This has become one of the motivations for the multiverse scenario.
7. The modern use of the term *quintessence* was introduced in R. R. Caldwell, R. Davé, and P. J. Steinhardt (1998). Cosmological imprint of an energy component with general equation-of-state, *Phys. Rev. Lett.* **80**, 1582–1585.

8. The energy density of radiation decreases more rapidly with expansion than does that of cold pressureless matter. This is because the number of photons is conserved, so the number density decreases inversely with volume, $1/a^3$. But because the photons are also redshifted with the expansion of the Universe, the energy of each photon also decreases as $1/a$. Thus $\rho \approx 1/a^4$.
9. From the general relationship $\rho \sim a^{-3(1+w)}$, we see that if w is smaller than -1 the dark energy density actually increases as the Universe expands – the case of phantom dark energy leading to a "big rip."
10. There are various speculative mechanisms for the local, but not cosmological, suppression of fifth forces, as described in Chapter 10.
11. In a sense, this does not differ from scalar fields, which would also be expected to modify the gravitational attraction by adding new variables – new degrees of freedom.
12. There are, for example, higher-dimensional theories which add a term to the Friedmann equation that results in late-time acceleration of expansion. See, for example, the theory of G. R. Dvali, G. Gabadadze, and M. Porrati: (2000). 4D gravity on a brane in 5D Minkowski space, *Phys. Lett. B* **485**, 208.
13. The degeneracy can be lifted in strong lenses if the background source, usually an active galactic nucleus, is intrinsically variable. The variability in one image is tracked by that in other images with a time delay due to the different path lengths (as well as the different gravitational field that is probed). This can provide a less ambiguous determination of the angular size distances.
14. S. D. M. White raised this question several years ago in a widely discussed preprint: *Fundamentalist physics: why dark energy is bad for astronomy* (arXiv:0704.2291). He emphasized that the problem of dark energy is of interest to two communities – astronomers and high-energy physicists, but these two communities have different methodologies and different scientific cultures. Physicists routinely invest enormous resources in experiments with a limited but significant purpose (the LHC, for example), while astronomers more typically invest in multipurpose instrumentation (the Hubble Space Telescope, for example) that is of interest to a diverse community. White argues that, while dark energy is undoubtedly an important observational problem, the large investment in resources and personnel may not lead to significant progress and may, in fact, limit research in other interesting directions. The question I pose here is similar: is it possible to reveal the nature of dark energy through astronomical observations, even with unlimited investment?

7 Dark Matter

1. C. Jones and W. Forman (1984). The structure of clusters of galaxies observed with Einstein, *Astrophys. J.* **276**, 38–55.
2. J. P. Kneib, R. S. Ellis, I. Smail, W. J. Couch, and R. M. Sharples (1996). Hubble Space Telescope observations of the lensing cluster Abell 2218, *Astrophys. J.* **471**, 643–656
3. For relatively nearby lenses the dependence on the cosmological model is secondary to the dependence on the lens mass.
4. Essentially, the observations imply that any alternative theory of gravity must reproduce the deflection of light equivalent to that provided by general relativity plus the dark mass implied by Newtonian dynamics. This requirement is quite restrictive and not generally satisfied by alternative theories such as a scalar–tensor theory with a conformal coupling to the Einstein metric. See J. D. Bekenstein and R. H. Sanders (1994). Gravitational lenses and unconventional gravity theories, *Astrophys. J.* **429**, 480–490.
5. The rotation curves shown here are based on observations done at the Westerbork Synthesis Radio Telescope (WSRT) and published by K. Begeman: (1987). HI rotation curves of spiral galaxies, *Astron. Astrophys.* **223**, 47–60.
6. This result was discovered in the 1970s and 1980s, primarily through the use of radio telescopes. The pioneering radio astronomers were M. S. Roberts and R. N. Whitehurst, G. S. Shostak and D. H. Rogstad, Albert Bosma, and, at optical wavelengths, V. C. Rubin, W. K Ford and N. Thonnard. This is discussed at length in my book, *The Dark Matter Problem: A Historical Perspective*, where a complete list of references is given.

7. The important paper describing the results of cosmic N-body simulations demonstrating the development of CDM halos is that of J. Navarro, C. Frenk, and S. White: (1996). *Astrophys. J.* **463**, 563–575. The characteristic form for the density distribution in the NFW halo (the density of dark matter as a function of radius) is

$$\rho = \frac{\rho_0}{(r/R_s)[1+(r/R_s)]^2}.$$

The model has two adjustable parameters, ρ_0 and R_s, that set the mass (or velocity) and size scales of the halo.
8. W. J. G. de Blok (2010). The core-cusp problem, *Advances in Astronomy* **2010**, 789293; see also arXiv:0910.3538.
9. A. Klypin, A. Kravtsov, O. Valenzuela, and F. Prada (1999). Where are the missing galactic satellites?, *Astrophys. J.* **522**, 82–92.
10. See the results on galaxy–galaxy lensing described by F. Brimioulle, S. Seitz, M. Lerchster, R. Bender, and J. Snigula (2013). *Mon. Not. R. Astron. Soc.* **432**, 1046–1102.
11. The "bullet" cluster is the prototype and most famous example of such a process. Here, in this apparent collision between two clusters, X-ray observations reveal the collisional gaseous component being left behind, while the two dissipationless components, stars and dark matter, as revealed by optical observations and lensing, pass on through (see D. Clowe, et al. (2006). *Astrophys. J.* **648**, L109). Presumably, the same could happen with protogalaxies.
12. The identity of the lowest-mass superpartner is, of course, unknown, but a possible candidate would be the neutralino, which is a composite particle – a mixture of a photino, a zino and a higgsino (super particles of the photon, the Z boson and the Higgs particle).
13. A useful review of this subject is given in Stefano Profumo (2015). TASI 2012 lectures on astrophysical probes of dark matter (arXiv:1301.0952). In particular, the point is made that this coincidence may not be so miraculous.
14. M. Ackermann, et al. (The Fermi-LAT Collaboration) (2015). Searching for dark matter annihilation from Milky Way dwarf spheroidal galaxies with six years of Fermi Large Area telescope data, *Phys. Rev. Lett.* **115**, 231301.
15. D. Hooper and L. Goodenough (2009). Dark matter annihilation in the Galactic Center as seen by the Fermi Gamma Ray Space Telescope, *Phys. Lett. B* **697**, 412–428.
16. In 2014 and 2015 there has been considerable interest in an apparent spectral line detected in individual and stacked X-ray observations of galaxy clusters, the Galactic Center and possibly the Andromeda Galaxy at X-rays – the 3.5-keV line. This could be due to the decay of a dark matter particle with a mass of 7 keV, possibly a sterile neutrino, with a lifetime of roughly 7×10^{27} seconds (10 million Hubble times). However, the line could also be due to potassium that has lost 17 of its electrons, depending quite critically upon the abundance of potassium in the astrophysical sites. Tesla Jeltema and Stafano Profumo (arXiv:1512.01239) have recently demonstrated that the line is not observed in the Draco dwarf spheroidal galaxy, supposedly a system that is heavily dominated by dark matter. The constraint on the lifetime of the hypothetical dark matter problem is greater than 3×10^{28} seconds, four times longer than that implied by the detection in galaxy clusters; the line does not result from dark matter decay. In general, every such announcement generates considerable excitement over the possibility of discovery of the dark matter. This particular incident demonstrates again that such claims are not credible until there is consistency with other results, either with astronomical observations or, preferentially, direct detection in terrestrial laboratories.
17. M. Aguilar, et al. [AMS Collaboration] (2013). AMS-02 provides a precise measure of cosmic rays, *CERN Courier* **53**(8), 23.
18. A very complete discussion of these effects is given in M. De Mauro, F. Donato, N. Fornego, and A. Vittino (2015). Improved constraints on the acceleration history of the Universe and the properties of the dark energy (arXiv:1507.0700).
19. M. G. Aartsen, et al. (2014). Observation of high-energy astrophysical neutrinos in three years of IceCube data, *Phys. Rev. Lett.* **113**, 101101.
20. The nuclear reactions occurring in the solar interior that convert hydrogen into helium also produce neutrinos, but these are low-energy particles – only up to 400 keV.

21. E. Aprile, et al. (2012). Dark matter results from 225 live days of XENON100 data, *Phys. Rev. Lett.* **109**, 181301.
22. D. S. Abott, et al. (2014). First results from the LUX Dark Matter Experiment at the Sanford Underground Research Facility, *Phys. Rev. Lett.* **112**, 091303.
23. For example, if the result on a possible detection of an 8.6-GeV WIMP were correct, LUX should have seen more than 1500 events during its 85-day run. It saw none that could be attributed to dark matter–nucleon scattering.
24. There are, of course, a number of phenomena that exhibit an annual modulation: for example, the hours of daylight, the ambient temperature at temperate latitudes, and, of possible relevance here, the flux of atmospheric muons that correlates with the density of air in the upper atmosphere. However, several studies have shown that the muon modulation cannot explain the DAMA result (see J. Klinger and V. A. Kudryavtsev (2015). Muon-induced neutrons do not explain the DAMA data, *Phys. Rev. Lett.* **114**, 151301 (arXiv:1503.07225 [hep-ph])).

8 MOND

1. B. Famaey and S. McGaugh (2012). Modified Newtonian dynamics (MOND): Observational phenomenology and relativistic extensions, *Living Rev. Relativity* **15**, 10.
2. As described in the previous chapter, the density distribution in dark matter halos that emerge from cosmic N-body calculations have a different form: at large radii the density declines as $1/r^3$ which should lead to a slowly declining rotation curve. This is not observed.
3. Fritz Zwicky, the discoverer of the mass discrepancy in clusters of galaxies and the first to propose dark matter, supposedly referred to his Mt. Wilson colleagues as "spherical bastards" – no matter which way you look at them, they are bastards.
4. Emmy Noether was one of the greatest mathematicians of the twentieth century and was generally recognized as such by contemporaries such as Albert Einstein and Herman Weyl. Her achievements are particularly noteworthy in the context of her time and place: the barriers to women in such professions were almost insurmountable.
5. This symmetry (also called Lorentz covariance) is named for the Leiden professor Hendrik Lorentz, who in the early years of the twentieth century wrote down the mathematical rules for transforming from one moving frame to another – the Lorentz transformations that leave the speed of light unchanged. These transformation rules allowed Einstein to formulate the Special Theory of Relativity.
6. M. Milgrom (2009). The MOND limit from space–time scale invariance, *Astrophys. J.* **698**, 1630–1638.
7. The Newtonian gravitational acceleration about a point mass M is given by $g_N = GM/r^2$. Multiplying distance and time by a factor b, we find that the right-hand side of this equation scales as $1/b$ but the right-hand side as $1/b^2$; i.e., the equation is not invariant to the value of b.
8. That is, $g = GM/(rr_0)$, where both sides scale as $1/b$; i.e., b does not appear in the relation.
9. R. Sanders (1996). The published extended rotation curves of spiral galaxies: Confrontation with modified dynamics, *Astrophys. J.* **473**, 117–129.
10. R. A. Swaters, R. H. Sanders, and S. S. McGaugh (2010). Testing modified Newtonian dynamics with rotation curves of dwarf and low surface brightness galaxies, *Astrophys. J.* **718**, 380–391.
11. R. Sancisi (2004). "The visible matter – dark matter coupling" S. D. Ryder, D. J. Pisano, and K. C. Freeman (eds.) *Dark Matter in Galaxies, Proc. IAU Symposium* 220, pp. 233–40, San Francisco: Astronomical Society of the Pacific, 2004.
12. S. S. McGaugh (2012). The baryonic Tully–Fisher relation of gas-rich galaxies as a test of CDM and MOND, *Astron. J.* **143**, 40.
13. See F. Brimioulle, cited in note 10, Chapter 7.
14. M. Milgrom (2013). Testing the MOND paradigm of modified dynamics with galaxy–galaxy gravitational lensing. *Phys. Rev. Lett.* **111**, 041105.
15. See, for example, R. Ibata, C. Nipoti, A. Sollima, M. Bellazzini, S. C. Chapman, and E. Dalessandro (2013). Do globular clusters possess dark matter halos? A case study in NGC 2419, *Mon. Not. R. Astron. Soc.* **428**, 3648–3659.

16. M. W. Auger, et al. (2010). The Sloan Lens ACS Survey. X. Stellar, dynamical, and total mass correlations of massive early-type galaxies, *Astrophys. J.* **724**, 511–525.
17. R. H. Sanders (2014). A dearth of dark matter in strong gravitational lenses, *Mon. Not. R. Astron. Soc.* **439**, 1781–1786.
18. In the first case we might expect a_0 to vary with cosmic time as H_0; in the second case the expectation would be that a_0 is constant, as is the cosmological term Λ.
19. W. G. Unruh (1976). A note on black hole evaporation, *Phys. Rev. D* **14**, 870–892.
20. See G. W. Gibbons and S. W. Hawking (1977). Cosmological event horizons, thermodynamics, and particle creation, *Phys. Rev. D* **15**, 2738–2751. In the case of the Universe, with $a_0 \approx 10^{-10}$, the temperature of de Sitter space would be rather frigid: $T = 10^{-22} K$.
21. M. Milgrom (1998). The modified dynamics as a vacuum effect (arXiv: astro-ph/9805346).
22. See, for example, R. H. Sanders (2003). Clusters of galaxies with modified Newtonian dynamics, *Mon. Not. R. Astron. Soc.* **342**, 901–908.
23. R. H. Sanders (2007). Neutrinos as cluster dark matter, *Mon. Not. R. Astron. Soc.* **380**, 331–338.
24. G. W. Angus (2009). Is an 11 eV sterile neutrino consistent with clusters, the cosmic microwave background and modified Newtonian dynamics? *Mon. Not. R. Astron. Soc.* **394**, 527–532.
25. M. Milgrom (2008), Marriage à-la-MOND: Baryonic dark matter in galaxy clusters and the cooling flow puzzle, *New Astron. Rev.* **51**, 906–915.
26. M. Milgrom (2015). Ultra-diffuse cluster galaxies as key to the MOND cluster conundrum, *Mon. Not. R. Astron. Soc.* **454**, 3810–3815.
27. J. D. Bekenstein and M. Milgrom (2004). Does the missing mass problem signal the breakdown of Newtonian gravity?, *Astrophys. J.* **286**, 7–14.
28. J. E. Felten (1984). Milgrom's revision of Newton's laws - dynamical and cosmological consequences, *Astrophys. J.* **286**, 3–6; R. H. Sanders (1998). Cosmology with modified Newtonian dynamics, *Mon. Not. R. Astron. Soc.*, **296**, 1009–1018.
29. J. D. Bekenstein (2004). Relativistic gravitation theory for the modified Newtonian dynamics paradigm, *Phys. Rev. D* **70**, 083509. I proposed an earlier version of this sort of relativistic theory for MOND (R. H. Sanders (1997). A stratified framework for scalar-tensor theories of modified dynamics, *Astrophys. J.* **480**, 492–502), building on earlier ideas of Bekenstein on disformal coupling of a scalar field to the Einstein metric. The problem with my theory was that the vector field was not dynamical; it acted on matter but was not acted upon by matter – an anathema in the post-Einsteinian world. Bekenstein corrected this problem in his theory, where all fields are dynamical and have their own sources.

9 Dark Matter, MOND and Cosmology

1. We should note that there are difficulties for the standard paradigm that go beyond individual galaxies. Pavel Kroupa and his collaborators have pointed out that there are problems implied by the spatial distribution of dwarf galaxies in the neighborhood of the Milky Way: many of these dwarfs are not randomly distributed in space but appear to lie in a great circle about the Galaxy. In the context of the CDM galaxy formation scenario, this is only possible if these objects are not primordial but have formed via an interaction process, e.g., a tidal disruption process of a larger galaxy or galaxies in orbit about the Milky Way. The internal kinematics of these objects should reflect this different formation history; they should be generally void of dark matter, but they are not. They also lie on the same Tully–Fisher relation as all other galaxies. Kroupa argues that this, in effect, falsifies the standard view of galaxy formation within the CDM cosmogony. See P. Kroupa (2012). The dark matter crisis: falsification of the current standard model of cosmology, *Pub. Astron. Soc. Aust.* **29**, 395–433.
2. A beam of neutrinos is actually a mixture of all types. Because the neutrinos have different masses, they propagate at different velocities; viewed quantum mechanically, this results in an interference of the wave pattern in the beam, first to predominately one type and then to another. This was first detected by the Super-Kamiokande experiment in Japan in 1998, for neutrinos produced in atmospheric decay of muons over energies ranging from hundreds of MeV up to one TeV and with a baseline of the diameter of the Earth.

3. To form the halos of normal spiral galaxies such as the Milky Way, the neutrino mass would have to be greater than 20 or 30 eV. See the classic paper by S. Tremaine and J. E. Gunn (1979). Dynamical role of light neutral leptons in cosmology, *Phys. Rev. Lett.* **42**, 407–410.
4. The idea that long-wavelength scalar field oscillations might comprise the dark matter is not new; for example, see W. H. Press, B. S. Ryden, and D. N. Spergel (1990). Single mechanism for generating large-scale structure and providing dark missing matter, *Phys. Rev. Lett.* **65**, 1084–1087.
5. R. H. Sanders (2005). A tensor–vector–scalar framework for modified dynamics and cosmic dark matter, *Mon. Not. R. Astron. Soc.* **363**, 459–468.
6. L. Berezhiani and J. Khoury (2015). Theory of dark matter superfluidity, *Phys. Rev. D* **92**, 103510.
7. M. Milgrom (2009). Bimetric MOND gravity, *Phys. Rev. D* **80**, 123536.
8. The term "brane" comes from "membrane." The idea is that the Universe is a four-dimensional sheet in a higher (possibly five-dimensional) world. There may be another such brane separated by a few millimeters in the fifth dimension. The idea that there is another entire universe a few millimeters away from our own has a certain appeal.
9. L. Bernard and L. Blanchet (2015). Phenomenology of dark matter via a bimetric extension of general relativity, *Phys. Rev. D* **91**, 103536.
10. T. Jacobson and D. Mattingly (2000). Gravity with a dynamical preferred frame, *Phys. Rev. D* **64**, 024028. For a version of such theories leading to MOND see T. G. Zlosnik, P. G. Ferreira, and G. D. Starkman (2007). Modifying gravity with the aether: An alternative to dark matter, *Phys. Rev. D* **75**, 044017.
11. D. Blas, O. Pujolas, and S. Sibiryakov (2010). A healthy extension of Horava gravity, *Phys. Rev. Lett.* **104**, 181302.
12. L. Blanchet and S. Marsat (2011). Modified gravity approach based on a preferred time foliation, *Phys. Rev. D* **84**, 044056; R. H. Sanders (2011). Hiding Lorentz invariance violation with MOND, *Phys. Rev. D* **84**, 084024.
13. E. Verlinde (2011). On the origin of gravity and the laws of Newton, *J. High Energy Phys.* **2011**(04), 029 (arXiv:1001.0785).
14. See, for example, the blog of Luboŝ Motl (http://motls.blogspot.nl/2010/01/erik-verlinde-why-gravity-cant-be.html) for a heated discussion of why gravity, as a reversible force, cannot be entropic. This is particularly interesting if one thinks that physicists are unemotional, Mr. Spock-like beings who deal with one another as cool overly-polite automatons. The blog is a glimpse into the true culture of theoretical physicists.
15. For example, see P. V. Pilkhitsa (2010). MOND reveals the thermodynamics of gravity (arXiv:1010.0318).
16. S. Dodelson and M. Liguori (2006). Can cosmic structure form without dark matter? *Phys. Rev. Lett.* **97**, 231301.
17. C. Skordis, D. F. Mota, P. G. Ferreira, and C. Boehm (2007). Large scale structure in Bekenstein's theory of relativistic modified Newtonian dynamics, *Phys. Rev. Lett.* **96**, 011301.
18. G. W. Angus, A. Diaferio, B. Famaey, and K. J. van der Heyden (2013). Cosmological simulations in MOND: the cluster scale halo mass function with light sterile neutrinos, *Mon. Not. R. Astron. Soc.* **436**, 202–211.
19. See Zlosnik, Ferreira, and Starkman (2007) cited in note 10 above.
20. Concerning initial conditions, inflation by itself cannot obviously explain all special properties of the Universe. For example, there is the baryon abundance: this is related to the asymmetry between matter and anti-matter – the fact that the Universe is not totally devoid of baryonic matter as might be expected if there were, in the early Universe, equal densities of baryons and anti-baryons. The photon-to-baryon ratio reveals the early dominance of matter over anti-matter to be about one part in 10^9. This is not understood but is probably due to non-conservation of the baryon number and may be an aspect of physics beyond the standard model. It is curious that the result yields a baryon abundance of about $\Omega_b \approx 0.05$ – within a factor of five of the dark matter abundance which, supposedly, results from different physics.
21. It might be argued that the pattern of acoustic peaks in the power spectrum of the CMB is predictive because the theory was worked out before the observations were made. However,

given the basic mechanism of phase-focussed acoustic oscillations, the basic pattern will be present. No one predicted the details of the location and amplitude of the peaks; this requires fitting of at least six adjustable parameters having degeneracies with additional parameters. MOND, on the other hand, accurately *predicts* more than 100 rotation curves with one new fixed constant. On this basis, it can be argued that MOND is epistemologically superior (forgetting about the reductionist preference for cosmology over individual objects).

Index

a_0, MOND acceleration constant, 101, 102, 105, 106, 108, 110–113, 116, 120, 122, 123

Aartsen, M., 136
Abott, D., 137
Ackermann, A., 136
acoustic oscillations, 42, 43, 45–47, 51
   baryon abundance, 45, 51, 53, 55, 64, 82
   dark matter abundance, 45
   phase focussing, 46
Ade, P., 133
adiabatic fluctuations, 24, 63
Albrecht, A., 46, 133
Alexander, S., 134
Alpher, R., 20
AMS, Alpha Magnetic Spectrometer, 86, 87
Anderson, P. W., 130
Angus, G., 118, 138, 139
anthropic principle, 10
Aprile, E., 137
astration, 37
Auger, M., 138
axis of evil, 61

baryon acoustic oscillations, 58, 59, 76
baryonic matter, 3, 7, 37–39, 41, 42, 44, 45, 49, 51–53, 55, 57, 59, 70, 72, 77, 78, 81–84, 87, 99, 101, 106, 112, 113, 116, 117, 119, 120, 124, 131
Begeman, K., 135
Bekenstein, J., 112, 113, 121, 123, 135, 138
Bekenstein–Milgrom non-relativistic theory, 112
Bellazzini, M., 137
Bender, R., 136
Bennett, C., 132, 133
Bentley, R., 13, 131
Berezhiani, L., 120, 139
Bernard, L., 120, 139
Bethe, H., 19–21
bi-scalar gravity, cosmic dark matter and MOND, 120
Bieri, L., 131

Big Bang model, 4, 7, 8, 17, 19–23, 25, 26, 28–31, 36, 37, 44, 49, 53, 55, 66
big rip, 57, 72, 76
bimetric theory, 120, 123
Binachi, E., 130
Biswis, T., 134
Blanchet, L., 120, 139
Blas, D., 139
Boehm, C., 139
Bondi, H., 20
BOOMERANG, 41
Bose–Einstein condensate, as dark halo, 120
Bosma, A., 135
bosons, 40, 71, 72, 119
   as dark matter, 119
   Higgs, 92
Branch, D., 134
brane-world, 120, 139
Brimioulle, F., 136
Burbidge, E., 21, 132
Burbidge, G., 21, 132

Caldwell, R., 134
Carter, B., 131
Casertano, S., 134
CDM, cold dark matter, 2, 3, 37, 40, 41, 52, 57, 72, 80–82, 86
CDMS, Cryogenic Dark Matter Search, 90, 91, 94, 95
CERN, European Organization for Nuclear Research, 91
Chandrasekhar limit, 68
Chandrasekhar, S., 68
Chapman, S., 137
Cheng, E., 132
Chornock, R., 134
Clowe, D., 136
clusters of galaxies
   dark matter, 39, 40, 80, 82, 83, 112
   MOND, 111
   the bullet, 83, 112

141

CMB, cosmic microwave background radiation, 1, 8, 22–25, 29, 33, 35–37, 132
  angular power spectrum, 47, 48
  anisotropy, 23, 26, 38, 41, 43, 47, 49, 51, 57, 58, 60, 61, 63, 64, 70, 74, 82, 115–119, 122, 123
  dipole anisotropy, 24
COBE, Cosmic Background Explorer, 22, 23, 25, 26, 41
Coc, A., 133
concordance model, 49
Copernicus, 61
Copi, C., 134
Core–CUSP problem, 81
cosmic rays, anti-matter to matter ratio, 86
Couch, W., 135
Coulson, D., 133

Dalessandro, E., 137
Daly, R., 134
DAMA-LIBRA, dark matter modulation experiment, 91, 95
dark energy, 2, 3, 11, 41, 45, 47, 52, 53, 55–58, 61, 64–67, 69–77, 81, 82, 113, 114, 116, 119, 120, 124, 134
dark matter, 1–5, 11, 39–42, 44, 45, 47–49, 52, 53, 56, 57, 59, 60, 64, 70, 72, 76–94, 96, 97, 102–104, 107, 108, 112–124, 136, 137, 139
  direct detection, 89
  indirect detection, 85
Davé, R., 134
de Bernardis, P., 133
de Blok, W., 136
de Broglie wavelength, 119
De Mauro, M., 136
de Sitter space radiation, 110
de Sitter universe, 15, 18, 20, 30–33, 41, 47, 48, 64, 66, 67, 75, 110, 111
de Sitter, W., 14, 17
decoupling, recombination of primordial plasma, 9, 38, 42, 46, 49, 50, 52, 58, 59, 70, 74, 81, 84, 116, 118, 119, 123
degenerate pressure of fermions, 68
deuterium abundance as a baryometer, 37
Diaferio, A., 139
Dicke, R., 23, 28, 132
dipolar dark matter, 120
Dodelson, S., 133, 139
Donato, F., 136
Dvali, G., 135

Eddington, A., 17, 30
Einstein–aether theories, 121–123
Einstein, A., 3, 5, 8, 14, 17–19, 31, 33, 41, 74, 116, 121, 132
Einstein field equation, 2, 14, 47, 70, 74, 75
Einstein ring, 108, 109
Eisenstaedt, J., 132
Eisenstein, D., 60, 134

Ellis, R., 135
emergence, 9, 130
Enuma Elish, Sumerian creation myth, 6
equation of state parameter w, 52, 55, 57, 71, 74, 76
equivalence of moving frames, 121
equivalence principle, 74
Evrard, A., 133

Famaey, B., 97, 137, 139
Felten, J., 138
Ferguson, H., 134
Fermi exclusion principle, 118
Fermi Gamma-ray Space Telescope, 85, 86, 89
Ferreira, P., 133, 139
fifth force, 74
Filippenko, A., 134
fine-tuning, 10, 28, 29, 36, 106
fine-tuning of initial conditions, 28
Fixsen, D., 132
fluctuations
  adiabatic, 34
  scalar, tensor, 34
Ford, K., 131
Ford, W., 135
Forman, W., 135
Fornego, A., 136
Fowler, W., 21, 37, 132
Freeman, K., 137
Frenk, C., 133, 136
Friedmann, Alexander, 14, 17, 18
Friedmann models, 15, 18, 23
Frieman, J., 134

Gabadadze, G., 135
Gales, J., 132
Galileo Galilei, 12, 74
Gamow, G., 19, 20, 23
Gibbons, G., 138
Gliner, E., 31
globular star clusters, 16, 21, 108
Gold, T., 20
Goodenough, L., 86, 136
Gottl, S., 134
Grand Unification, 32, 35
gravitational collapse, 7, 23–27, 112
gravitational lensing, 74, 76, 78, 82, 106, 108, 112, 113
Gunn, J., 139
Guth, A., 32, 33, 132

Hamilton, E., 130
Hanany, S., 133
harmonic peaks, support for inflation, 45
Hawking radiation, 110
Hawking, S., 138
helium abundance as a chronometer, 37
Herman, R., 20
Herschel, W., 13

Hertzsprung–Russell diagram, 21
Hess, S., 134
Hindu creation myth, 8
Hinshaw, G., 133
Hooper, D., 86, 136
horizon, 25, 29, 34, 41, 42, 44–46, 58, 59, 61, 63, 70, 110, 122
Hoyle, F., 8, 20–22, 37, 132
Hu, W., 133, 134
Hubble constant, 17, 20, 21, 28, 36, 49, 52, 53, 57, 58
Hubble diagram, 1, 17, 56, 64, 68, 69, 82, 114, 116, 124
Hubble, E., 16, 131
Hubble radius, 23
Huterer, D., 134

Ibata, R., 137
IceCube neutrino telescope, 90
Illiev, I., 134
inflation, 28, 32–35
   horizon problem, 33
   impact on cosmology, 35
   quantum fluctuations, 34
inflaton, 33

Jacobson, T., 139
Jeans length, 24, 25
Jeans, J., 23
Jeltema, T., 136
Jha, S., 134
Jones, C., 135

Kant, I., 13
Kapteyn, J., 13
Kepler, J., 12
Khoury, J., 120, 139
khronon field, as cosmic time, 121
Klinger, J., 137
Klypin, A., 136
Knapp, G., 134
Knebe, A., 134
Kneib, J.-P., 135
Knox, A., 132
Kragh, H., 132
Kravtsov, A., 136
Kroupa, P., 138
Kudryavtsev, V., 137
Kuhn, T., 131

Lampeitl, H., 134
Land, K., 134
Large Hadron Collider (LHC), 91
Larson, D., 133
Laughlin, R., 130
Le Verrier, U., 12
Leavitt, H., 16
Lemaître, G., 17–20
Lerchster, M., 136

Liguori, M., 139
Linde, A., 8, 32, 35, 132
Lorentz covariance, 98
Lorentz, H., 98
Lundmark, K., 16
LUX, Large Underground Xenon Experiment, 90, 91, 95

Macri, L., 134
Magueijo, J., 133, 134
Marsat, S., 139
Mather, J., 132
Mattingly, D., 139
MAXIMA, 41
McGaugh, S., 97, 105, 137
Milgrom, M., 3, 100, 101, 112, 120, 130, 137–139
Milky Way Galaxy, 13, 16, 21, 25, 61, 64, 81, 86, 89, 108, 110
missing galaxy satellites, 81
MOND, modified Newtonian dynamics, 3–5, 96, 99, 101–108, 110–124, 139
Mota, D., 139
Motl, L., 139
multiverse, 8, 11, 35, 36, 71

Navarro, J., 133, 136
neutrino telescope, 87, 88
neutrinos, 37, 39, 52, 53, 55, 58, 83, 85, 87–89, 92, 112, 117, 118, 123, 138
   sterile, 112, 118, 136
Newton, I., 8, 12–14, 23, 98, 111, 115, 116, 121
Nipoti, C., 137
Noether, E., 98, 137
North, J., 131
Notari, A., 134
nucleosynthesis
   baryon abundance, 37, 39, 45, 51, 81
   origin of the chemical elements, 18–20, 22
   stellar, 21
Nussbaumer, H., 131

Ostriker, J., 49, 133

Page, L. A., 132
Peebles, P. J. E., 37, 132
Penzias, A., 22
Perlmutter, S., 133
Pilkhitsa, P., 139
Pisano, D., 137
Planck, M., 18
Planck satellite, 42, 43, 51–53, 57, 62, 118
Planck units, 30
Porrati, M., 135
Prada, F., 136
Press, W., 139
Profumo, S., 136
Pujolas, O., 139

Rapetti, G., 134
red giant, 21
redshift, definition, 16
reductionism, 9, 115, 130
Riess, A., 133, 134
Roberts, M., 135
Robertson, H., 18
Robertson–Walker metric, 18
Rogstad, D., 135
Rovelli, C., 130
Rubin, V., 135
Ryden, B., 139
Ryder, S., 137

Sachs, R., 24, 25, 132
Sakharov, A., 31
Sancisi, R., 137
Sanders, R., 103, 135, 137–139
SCDM, standard CDM cosmology, 40, 81
Schramm, D., 37
Schwarz, D., 134
Schwarzschild, K., 30
Schwarzschild, M., 21, 132
scientific method, 10, 12, 13, 27, 125, 131
Seitz, S., 136
Shafer, R., 132
Shapley, H., 15
Sharples, R., 135
Shostak, G., 135
Sibryakov, S., 139
Singh, S., 132
Skordis, C., 139
Slipher, V., 16
Sloan Digital Sky Survey, SDSS, 60
Smail, I., 135
Smeenk, C., 132
Smith, G., 130
Smoot, G., 132
Snigula, J., 136
Sollima, A., 137
space–time scale invariance, 100
Spergel, D., 139
Starkman, G., 134
Starobinsky, A., 31, 32
Steady State model, 20–23, 36, 132
Steigman, G., 133
Steinhardt, P., 49, 36, 133, 134
stellar evolution, 21
Strauss, M., 134
Sullivan, M., 134
Super-Kamiokande, 138
supernovae, 87
  type 1a, 40, 45, 51, 55–57, 64, 68–70, 75, 82, 110, 114, 116, 124

supersymmetry, superparticles, 35, 39, 40, 71, 84, 92
Swaters, R., 137

Taylor, R., 22
Tegmark, M., 134
teleology, 9, 10
TeVeS, tensor–vector–scalar theory, 138
Thonnard, N., 135
Ting, S., 86
Tolman, R., 30, 132
Tremaine, S., 139
Tully–Fisher relation, 101, 104–107, 113, 116
Turner, M., 37, 134
twin matter, 120

universal preferred frame, 122
Unruh radiation, 110
Unruh, W., 138
Uzan, J.-P., 133

vacuum, false and true, 33
Vaid, D., 134
Valenzuela, O., 136
van der Heyden, K., 139
Vangioni, E., 133
Verlinde, E., gravity as an entropic force., 122, 139
Vittino, A., 136
von Seeliger, H., 13
von Weizsäcker, Carl Friedrich, 19–21, 132

Wagoner, R., 37, 132
Walker, A., 18
Watson, W., 134
Weinberg, S., 134
Weinland, J., 133
Wheeler, J., 14, 131
white dwarf, 21, 68
White, S., 76, 133, 135, 136
Whitehurst, R., 135
Wilson, R., 22
WIMPs, weakly interacting massive particles, 84–86, 88, 89, 91, 94, 95
WMAP, Wilkinson Microwave Anisotropy Probe, 26, 41, 51–53, 57, 62
Wojtak, R., 134
Wolfe, A., 24, 25, 132
Wright, E., 132

X-ray 3.5-keV line, 136
XENON 100, 90, 91, 95

Yepes, G., 134

Zeldovich, Y., 31, 32
Zwicky, F., 77, 78, 137